Solvation
Thermodynamics

Solvation Thermodynamics

Arieh Ben-Naim

Department of Physical Chemistry
The Hebrew University of Jerusalem
Jerusalem, Israel

Plenum Press • New York and London

Randall Library UNC-W

Library of Congress Cataloging in Publication Data

Ben-Naim, Arieh, 1934–
   Solvation thermodynamics / Arieh Ben-Naim.
     p. cm.
   Includes bibliographical references and index.
   ISBN 0-306-42538-6
   1. Solvation. 2. Thermodynamics. 3. Solution (Chemistry) I. Title.
QD543.B36   1987
541.3′416 — dc19   87-21738
                                                             CIP

The following journals have granted permission to reproduce previously published figures:
American Journal of Physics (Figures 4.1–4.14), *Journal of Chemical Physics* (Figures 2.13, 2.15–2.22, 2.29, and 2.30), and *Journal of Physical Chemistry* (Figures 2.4–2.6, 2.24–2.28, and 2.39–2.41).

© 1987 Plenum Press, New York
A Division of Plenum Publishing Corporation
233 Spring Street, New York, N.Y. 10013

All rights reserved

No part of this book may be reproduced, stored in a retrieval system, or transmitted in any form or by any means, electronic, mechanical, photocopying, microfilming, recording, or otherwise, without written permission from the Publisher

Printed in the United States of America

QD
543
. B36
1987

Preface

This book deals with a subject that has been studied since the beginning of physical chemistry. Despite the thousands of articles and scores of books devoted to solvation thermodynamics, I feel that some fundamental and well-established concepts underlying the traditional approach to this subject are not satisfactory and need revision.

The main reason for this need is that solvation thermodynamics has traditionally been treated in the context of classical (macroscopic) thermodynamics alone. However, solvation is inherently a molecular process, dependent upon local rather than macroscopic properties of the system. Therefore, the starting point should be based on statistical mechanical methods.

For many years it has been believed that certain thermodynamic quantities, such as the standard free energy (or enthalpy or entropy) of solution, may be used as measures of the corresponding functions of solvation of a given solute in a given solvent. I first challenged this notion in a paper published in 1978 based on analysis at the molecular level.

During the past ten years, I have introduced several new quantities which, in my opinion, should replace the conventional measures of solvation thermodynamics. To avoid confusing the new quantities with those referred to conventionally in the literature as standard quantities of solvation, I called these "nonconventional," "generalized," and "local" standard quantities and attempted to point out the *advantages* of these new quantities over the conventional ones.

It was just very recently that I not only became aware that the newly proposed quantities have *advantages* over the conventional ones and could be applied in a uniform manner to a wider range of systems, but also became convinced that these proposed quantities are actually the ones that deserve to be referred to as *bona fide* measures of solvation thermodynamics. This conviction has compelled me to risk claiming the term *solvation thermodynamics* for the newly proposed quantities.

v

I realize that the usage of these terms with their newly assigned meanings will initially cause some confusion, since the reader's notions of solvation thermodynamics have been inevitably biased by exposure to the previously assigned meanings. However, I hope that this book will convince the reader willing to discard prejudices that the quantities used in this book are in fact the best ones for studying solvation thermodynamics. As in biology, I believe that in the long run the concept that is the fittest will survive.

By using the term *solvation* in the present approach, we do make a minor semantic sacrifice. The very term *solvation* implies the traditional distinction between a *solute* and a *solvent*. The former is being *solvated* by the latter. In the present approach we shall apply the term *solvation* in a more general sense. Instead of speaking about "a solute being solvated by a solvent," we shall be talking about "a molecule being solvated by a medium," i.e., we shall abandon the traditional distinction between a solute and a solvent. I believe, however, that the cost of this minor sacrifice is more than compensated by the enormous increase in generality afforded by the present treatment. For example, the present treatment permits us an exact description of the thermodynamics of solvation not only in the limit of infinite dilution of solute, but over the entire physically realizable range of concentrations. Thus we shall be speaking of the solvation of, say, argon in water as well as the solvation of argon in any mixture of argon and krypton, including the case of the solvation of argon in pure argon. We shall also find that, once we have properly defined the solvation process, we may describe the thermodynamics of solvation without reference to *any* standard states. This is a considerable improvement relative to the present situation in the field.

The book is organized in four parts. The first presents the subject matter in an elementary, sometimes phenomenological manner. The reader who is interested in getting a general idea of the method and scope of applications could satisfy himself by reading this part only. However, for a deeper understanding of the ideas employed, it is necessary to consult various sections in Chapter 3. The latter chapter contains a more detailed treatment of subjects discussed, and referred to, in Chapters 1 and 2.

Chapter 2 includes various applications to specific systems, ranging from hard spheres to protein solutions. The selection of systems was done without any pretense at being exhaustive or up to date. In fact, while writing this work it became clear to me that there is an almost unlimited number of systems that could fit into the framework of this book. Obviously, some arbitrary decisions had to be made in including some systems while rejecting others.

The last chapter on mixing and assimilation was added because it has direct bearings on the correct interpretation of various components of the chemical potential. I believe that some of the fundamental misconceptions regarding the free energy and entropy of mixing that are so ubiquitous in the literature have led to various misinterpretations in the traditional approach to solvation thermodynamics. Reading this chapter, though not essential to understanding the rest of the book, could be quite useful for the comprehension of the various ingredients that contribute to the chemical potential.

The book is addressed primarily to experimentalists in the field of solution chemistry. Although most of the results derived in this book were obtained from thermodynamic quantities, this is not true for *all* the systems, as will be clear from Sections 1.6, 2.17, and 3.12 dealing with ionic solutions. The reason is that the fundamental relationships between solvation thermodynamics and experimental quantities have their origin in statistical mechanics, not in thermodynamics. In a sense, the whole approach as presented in this book may be viewed as being a hybrid between a pure thermodynamic approach and a statistical mechanical approach. We rely from the very outset on statistical mechanics, yet it is not a pure statistical mechanical approach in the sense that we do not attempt to compute solvation quantities using first-principle techniques of statistical mechanics.

The book was conceived in Jerusalem, Israel. The first draft was written in La Plata, Argentina, the second draft was written in Manila, Philippines, and the final draft and typing were completed in Bethesda, Maryland. I would like to express my indebtedness to my colleagues and hosts, Dr. Raul Grigera in La Plata and Dr. Claro Llaguno in Manila, who provided the favorable atmosphere in which to write the book while visiting their countries.

Thanks are also due to Drs. R. Battino, S. Goldman, Y. Marcus, A. P. Minton and S. Nir for reading parts of the manuscript and offering helpful comments and suggestions.

Arieh Ben-Naim

Jerusalem, Israel

Contents

Chapter 3. Further Theoretical Background

Chapter 4. On Mixing and Assimilation

Solvation
Thermodynamics

Chapter 1

Elementary Background

1.1. THE FUNDAMENTAL EXPRESSION FOR THE CHEMICAL POTENTIAL

In thermodynamics the chemical potential (CP) of a component s in a system may be defined in various ways. The most common and useful definitions are

$$\mu_s = \left(\frac{\partial G}{\partial N_s}\right)_{P,T,\mathbf{N}'} = \left(\frac{\partial A}{\partial N_s}\right)_{T,V,\mathbf{N}'} \tag{1.1}$$

where G is the Gibbs energy of the system, A is the Helmholtz energy, N_s is the number of s molecules in the system, T is the absolute temperature, P is the pressure, V is the volume, and $\mathbf{N}$ designates the set of numbers $N_1, \cdots, N_c$, where N_i is the number of molecules of the ith species in the system. The set $\mathbf{N}'$ is the same as $\mathbf{N}$ but excludes the number N_s for the species s.

In thermodynamics the CP is usually defined *per mole* rather than *per molecule*. However, since our fundamental expression for the CP will be derived from statistical mechanics, it will be more convenient to define the CP, as well as other partial molecular quantities, per molecule rather than per mole. The conversion between the two quantities involves the Avogadro number $N_A = 6.022 \times 10^{23}$ mol^{-1}. However, when presenting values of the thermodynamic quantities of solvation in Chapter 2, they will all be *per mole* of the corresponding component. The first derivative on the right-hand side (rhs) of equation (1.1) is taken at constant temperature and pressure. These variables are the most easily controllable in actual experimental work. Hence, this definition of the CP is the more useful from the practical point of view. However, from the theoretical point of view it is somewhat easier to work with the second derivative on the rhs of equation (1.1). Here the temperature and volume are kept constant. The corresponding thermodynamic potential A is conveniently related to the canonical partition function of the system (see Section 3.1).

It is well known that thermodynamics alone does not provide the functional dependence of the CP on the temperature, pressure, or composition of the system. In principle, this is obtainable from statistical mechanics. Moreover, statistical mechanics is presumed to furnish an explicit dependence of μ_s on all the relevant molecular parameters pertinent to all the molecules present in the system.

In some very special cases a partial dependence of the CP on pressure or composition is handled within the framework of thermodynamics. For instance, if the system is an ideal gas then the CP of component s may be written as

$$\mu_s = \mu_s^{\circ g} + kT \ln P_s \qquad (1.2)$$

where P_s is the partial pressure of the component s in atmospheric units, k is the Boltzmann constant, and $\mu_s^{\circ g}$ is a constant independent of the partial pressure P_s.

A second case is a very dilute solution of a nonionic solute s in a solvent. In this case, Henry's law is observed and the CP of s may be expressed in the form

$$\mu_s = \mu_s^{\circ l} + kT \ln x_s \qquad (1.3)$$

where x_s is the mole fraction of s and $\mu_s^{\circ l}$ is referred to as the standard CP, a quantity independent of x_s.

Clearly, both equations (1.2) and (1.3), though useful, are valid for very restrictive cases (ideal gases and solutions). Most real systems studied in the laboratory are neither ideal gases nor ideal solutions.

Throughout this book we shall use a very general expression for the CP that is not accessible from thermodynamics. It is based on some simple statistical mechanical considerations. Here we shall present only a qualitative description of the various factors involved in this expression. A more detailed derivation is deferred to Section 3.1.

For simplicity, we consider a two-component system at some temperature T and pressure P with N_A and N_B the number of molecules of components A and B, respectively. The extensive character of the Gibbs energy enables the mathematical derivative in equation (1.1) to be replaced by a difference, namely

$$\mu_A = G(T, P, N_A + 1, N_B) - G(T, P, N_A, N_B) \qquad (1.4)$$

i.e., the CP of component A is the change in the Gibbs energy caused by the addition of *one* A molecule to the system while keeping T, P, and N_B unchanged. This statement is valid for a macroscopic system, where the addition of *one* molecule may be viewed as an infinitesimal change in the variable N_A.

For the purpose of interpreting the various contributions to the CP, it is useful to introduce an auxiliary quantity which we shall refer to as the pseudochemical potential (PCP). This quantity is defined similarly to equation (1.4) but with the additional restriction that the center of mass of the newly added molecule be placed at a *fixed* position, say $\mathbf{R}_0$, within the system. We thus define the PCP of component A as

$$\mu_A^* = G(T, P, N_A + 1, N_B; \mathbf{R}_0) - G(T, P, N_A, N_B) \qquad (1.5)$$

Clearly, the difference between the definitions of μ_A and μ_A^* is only in the constraint imposed on the location of the center of mass of the newly added molecule.

We note that since our system is presumed to be macroscopic and homogeneous (no external fields), all the points in the system should be equivalent (except for a small region near the surface of the system that may be neglected for macroscopically large systems). Therefore, we use the notation μ_A^* rather than the more explicit one $\mu_A^*(\mathbf{R}_0)$ to stress the fact that μ_s^* is independent of the particular choice of $\mathbf{R}_0$. The asterisk is sufficient to indicate that we are dealing with a PCP and that we are constrained to a fixed point; the exact location of the selected point $\mathbf{R}_0$ is of no relevance to any of the following considerations.

Classical statistical mechanics provides us with a very useful relation between the CP, μ_s of a molecule s, and the corresponding PCP, μ_s^* of the same molecule in the same system. This relation is

$$\mu_s = \mu_s^* + kT \ln \rho_s \Lambda_s^3 \qquad (1.6)$$

A full derivation of this relation is given in Section 3.1. In equation (1.6), ρ_s is the number density of the component s, $\rho_s = N_s/V$, and Λ_s^3 is referred to as the momentum partition function. The product $\rho_s \Lambda_s^3$ is a dimensionless quantity. The applicability of classical statistical mechanics for our systems is based on the assumption that $\rho_s \Lambda_s^3 \ll 1$. Indeed, for most systems of interest in solution chemistry, this condition is fulfilled.[†]

[†] Liquid water might be an exception. At present it is not clear to what extent the translational degrees of freedom of a water molecule are separable and can be treated classically.

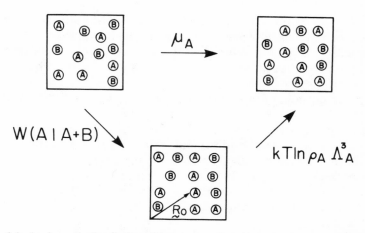

Figure 1.1. A schematic description of the process of adding one simple spherical molecule A to a mixture of A and B. First, the center of mass of the particle is placed at a fixed position in the system, and then the particle is released to wander in the entire volume. The corresponding contributions to the chemical potential of A are indicated next to each arrow.

Equation (1.6) for the chemical potential may be interpreted as follows. The chemical potential is a measure of the change in the Gibbs energy caused by the addition of one s particle to the system (at constant P, T, and composition). The split of μ_s into two parts corresponds to a splitting of the process of adding the particle s in two steps. First, we place the molecule at a *fixed* position, say $\mathbf{R}_0$, and the corresponding change in the Gibbs energy is μ_s^*. Next, we release the constraint imposed on the fixed position; this results in an additional change in the Gibbs energy, $kT \ln \rho_s \Lambda_s^3$. Since we are dealing with classical systems,[†] $\rho_s \Lambda_s^3 \ll 1$, the quantity $kT \ln \rho_s \Lambda_s^3$ is always negative. This quantity has been referred to as liberation Gibbs energy.[(1)‡] The two steps are described schematically in Figure 1.1 for a two-component system of A and B.

[†] More precisely, we should refer to systems for which the classical limit of statistical mechanics is applicable.

[‡] This term could also be referred to as the "translational Gibbs energy." However, there are cases where there exists no proper translation, yet an analogue of the liberation Gibbs energy term in the chemical potential does appear. Examples are discussed in Sections 3.16 and 3.17. In these cases the term $kT \ln \rho \Lambda^3$ is replaced by a term $kT \ln N/M$, where M is the total number of available sites. The latter term may be referred to as liberation Gibbs energy, but not translational Gibbs energy, since no proper translation exists in these systems. In this sense the term "liberation" has more general applicability, and the "translational Gibbs energy" is only a particular example of the "liberation Gibbs energy."

Although we did not offer a proof of relation (1.6) (see, however, Section 3.1), it is helpful to obtain a qualitative idea of the significance of the various factors that contribute to the liberation term $kT \ln \rho_s \Lambda_s^3$. When we release the particle that was constrained to a fixed position, it acquires translational kinetic energy; the corresponding contribution to the Gibbs energy of the releasing process is $kT \ln \Lambda_s^3$. Furthermore, once released, the particle may wander throughout the entire volume V of the system, giving rise to the term $-kT \ln V$. Finally, the particle that is at a fixed position is distinguishable from all other s particles in the system. Once it is released, it is assimilated[†] by all the N_s particles and therefore loses its distinguishability. This, together with the previous contributions, gives rise to the term $kT \ln \rho_s \Lambda_s^3$.

The most important property of equation (1.6), which constitutes the cornerstone of the entire content of this book, is its generality. It is valid for any kind of molecule (argon, benzene, or a protein), in any fluid mixture (i.e., in a multicomponent system), and for any concentration of s (from very dilute systems up to pure liquid s). The only assumption that has been introduced to render equation (1.6) valid is the applicability of classical statistical mechanics. As noted above, we shall deal in this book only with systems that obey classical statistical mechanics.

In equation (1.6) we have defined ρ_s as N_s/V in the canonical ensemble, i.e., when N_s and V are fixed variables. However, this equation and the significance of the two terms as interpreted above hold true for any other ensemble. For instance, in the $T, P, \mathbf{N}$ ensemble, we should read $\rho_s = N/\langle V \rangle$, where $\langle V \rangle$ is the average volume of the system. In the $T, V, \boldsymbol{\mu}$ ensemble we should read $\rho_s = \langle N_s \rangle / V$, where $\langle N_s \rangle$ is the corresponding average number of s particles in the system (for more details, see Section 3.1).

In some special cases we may rewrite equation (1.6) in a somewhat more detailed form. The simplest case is an ideal gas. If the internal partition function (i.e., rotational, vibrational, electronic, and nuclear) is denoted by q_s, then equation (1.6) reduces to

$$\mu_s = -kT \ln q_s + kT \ln \rho_s \Lambda_s^3$$
$$= (-kT \ln kT q_s / \Lambda_s^3) + kT \ln P_s$$
$$= \mu_s^{og} + kT \ln P_s \qquad (1.7)$$

where in the last expression on the rhs of equation (1.7) we have also written the CP of s in the conventional thermodynamic form.

[†] The term assimilation will be discussed in more detail in Chapter 4. It is used here only to stress the fact that the term $kT \ln \rho_s \Lambda_s^3$ is *not* a mixing term, as is so often referred to in the literature.

Clearly, when there are no interactions among the particles the PCP is simply

$$\mu_s^* = - kT \ln q_s \qquad (1.8)$$

The liberation term is the same as in equation (1.6).

In a condensed phase the internal degrees of freedom may or may not be affected by the interaction of s with its surrounding molecules. First, we consider a simple structureless particle, say a hard sphere or argon. In this case we assume that $q_s = 1$ and hence equation (1.6) reduces to

$$\mu_s = W(s \mid l) + kT \ln \rho_s \Lambda_s^3 \qquad (1.9)$$

In equation (1.9) we use the notation $W(s \mid l)$ to designate the average Gibbs energy of interaction of s with its entire surroundings. This is also referred to as the coupling work of s to the system. A more explicit expression for $W(s \mid l)$ is derived in Section 3.1.

The more general case is when s does have internal degrees of freedom. If these are unaffected by the interactions, we may simply add the term $- kT \ln q_s$ to equation (1.9). However, it is expected that in most cases the internal degrees of freedom will be affected by the interactions. A simple example where an internal rotational degree of freedom is affected is discussed in Section 3.2. In the most general case we shall use relation (1.6), in which μ_s^* includes both the coupling work (actually an average coupling work) and whatever effect the surroundings have on the internal partition function of s. We shall never need to spell out the details of μ_s^*; however, one should keep in mind the various factors that might contribute to its value in a condensed phase.

1.2. DEFINITION OF THE SOLVATION PROCESS AND THE CORRESPONDING SOLVATION THERMODYNAMICS

The term "solvation" or its more specific predecessor "hydration" has been used in physical chemistry probably since the creation of this branch of science. In the *Encyclopaedia Britannica* we find the following description of this term: "When a solvent and a solute molecule link together with weak bonds, the process is called solvation."[2] In another, now classical monograph by Gurney, we find: "The interaction that takes place when an ion is introduced into a solvent is called the solvation of the ion."[3]

It is evident from the above quotations that "solvation" is a term that is associated with the *interaction* between a *solute* and a *solvent*. This quite general description leaves, however, a great deal of freedom when we attempt to translate the term "solvation" into a more precise definition. What exactly is the "process of interaction?" What are the phenomena that are manifested by the solvation process? Which molecules shall we consider as "solute" and which as "solvent?" What do we mean by entropy, enthalpy, volume, etc., of solvation?

No wonder the term "solvation" has been used in a multitude of meanings. At present there seems to be no universal agreement on the very definition of the solvation process. Instead, it has been commonly assumed that some standard thermodynamic process, e.g., the transfer of argon from the gaseous phase at 1 atm pressure to a solution at some specified concentration, could serve as a definition of the solvation process and hence provide the grounds for defining all the corresponding thermodynamic quantities. A close examination of this approach[1] reveals that the various "standard processess" indeed *include* the "solvation process," and the pertinent standard thermodynamic quantities *include* the thermodynamic quantities of solvation. However, we shall see in Section 1.4 that the traditional standard thermodynamic quantities are, in principle, inadequate measures of the solvation phenomenon.

We now define the solvation *process* of a molecule s in a fluid *l* as the process of transferring the molecule s from a *fixed* position in an ideal gas phase *g* into a fixed position in the fluid or liquid phase *l*. The process is carried out at constant temperature T and pressure P. Also, the composition of the system is unchanged.

When such a process is carried out, we shall say that the *molecule* s is being solvated by the liquid phase *l*. If s is a simple spherical molecule, it is sufficient to require that the center of the molecule be fixed. On the other hand, if s is a more complex molecule, such as n-alkane or a protein, we require that the center of mass of the molecule be at a fixed position. We note, also, that in complex molecules the geometrical location of the center of mass might change upon changing the conformation of the molecule. In such cases we need to distinguish between the process of solvation of the molecule at a particular conformation and an average solvation process over all possible conformations of the molecule. A detailed example is worked out in Section 3.2.

One could also define the solvation process as above, but at constant volume rather than constant pressure. The definition given above is the one which may be related more directly to experimental quantities. However, for some theoretical considerations it might be more convenient to treat the constant-volume solvation process. The relation between the two is discussed in Section 3.1.

In defining a particular solvation process of a molecule s, we must specify the temperature, the pressure, and the composition of the liquid phase. There is no restriction whatsoever on the concentration of s in the system. This may be very dilute s in l, in which case the term *solute* in its conventional sense might apply for s. It may be a concentrated solution of s in l, or even a pure liquid s. Clearly, in the latter cases the conventional sense of the term *solute* becomes inappropriate. However, what remains unchanged is the conceptual meaning of the term *solvation* as a measure of the interaction between s and its entire surroundings l. This point will be clarified further when we introduce the various thermodynamic quantities of solvation below. As we shall see in Section 1.4, in the limit of very dilute solutions some of the thermodynamic quantities of *solvation*, as defined above, coincide with the conventional quantities of solvation. This is not the case, however, for higher concentrations of s in l. While conventional thermodynamics cannot be applied to these systems, the new definition, along with the pertinent thermodynamic quantities, can be applied without any restrictions on the concentration of s. In this sense the new definition generalizes the concept of *solvation* beyond its traditional limits. In fact this generalization extends the applicability of the concept of solvation from one concentration to an infinite range of concentrations.

It will be useful to introduce at this point the concept of a *solvaton*. The solvaton s is that particular molecule s the solvation of which is studied. This term is introduced to stress the distinction between the molecule serving as our "test particle" and other molecules of the same species that might be in the surroundings of the solvaton. In Section 3.18 we shall also require that solvatons do not interact with each other, but that they do interact with the rest of the system in exactly the same manner as a regular molecule of the same species.

Once we have defined the *process of solvation*, we may proceed to introduce the corresponding thermodynamic quantitites. We shall henceforth talk about *solvation entropy, solvation energy, solvation volume*, and so on, meaning the change in the corresponding thermodynamic quantity associated with the solvation process as defined above.

First, and of foremost importance, is the Gibbs energy of solvation of s in l. This is defined as

$$\Delta G_s^* = \mu_s^{*l} - \mu_s^{*ig} \tag{1.10}$$

where μ_s^{*l} and μ_s^{*ig} are the PCP of s in the liquid and in an ideal-gas (ig) phase, respectively. From the definition of the PCP given in the previous section, it is clear that ΔG_s^* is the Gibbs energy change for transferring s from a *fixed* position in an ideal-gas phase into a *fixed* position in the liquid phase l.

Since in the ideal-gas phase we have

$$\mu_s^{*ig} = -kT \ln q_s \tag{1.11}$$

it follows that ΔG_s^*, as defined in equation (1.10), includes all the effects due to the interaction between s and its entire environment. This may sometimes be split into the interaction Gibbs energy and the effect of l on the internal degrees of freedom of s. Such a distinction is not easy to achieve from experimental data.

A relatively simple situation arises when the solvaton has no internal degrees of freedom or when these are effectively unaffected by the surroundings.[†] In these cases we may write

$$\mu_s^{*l} = W(s \mid l) - kT \ln q_s \tag{1.12}$$

where q_s is exactly the same as in equation (1.11). Hence the solvation Gibbs energy reduces to

$$\Delta G_s^* = W(s \mid l) \tag{1.13}$$

which is simply the coupling work of s to l. In Section 3.1 we derive the statistical mechanical expression for $W(s \mid l)$, which reads

$$W(s \mid l) = -kT \ln \langle \exp(-B_s/kT) \rangle \tag{1.14}$$

where B_s is the total binding energy of s to the entire system l at some specific configuration and $\langle \; \rangle$ signifies an ensemble average over all possible configurations of all the molecules in the system, excluding the solvaton s, which is the molecule s under observation. Expression (1.14) may be interpreted as the average Gibbs energy of interaction of s with l. This should be clearly distinguished from the average *interaction* energy between s and l (see Section 3.8).

At this point we introduce a notation for two limiting cases of the quantity ΔG_s^*, namely

$$\Delta G_s^{*\circ} = \lim_{\rho_s \to 0} \Delta G_s^* \tag{1.15}$$

and

$$\Delta G_s^{*p} = \lim_{\rho_s \to \rho_s^p} \Delta G_s^* \tag{1.16}$$

[†] This is strictly true only for hard spheres. For any real molecule, even for inert gases, some effect of the environment on q_s must occur. This is certainly true at very high solvent densities.

The first quantity is the low-density limit, i.e., when s is very dilute in l. This is the region where conventional thermodynamics is applicable. We shall see in Section 1.4 that, incidentally, the quantity $\Delta G_s^{*\circ}$ coincides numerically, but not conceptually, with one of the standard Gibbs energies of solution. This is true only for the Gibbs energy and not for other thermodynamic quantities.

The second limit is the case of pure s, i.e., when the solvaton s is being solvated only by s molecules. This case has not been, and cannot be, treated by purely thermodynamic considerations. Here, ρ_s^p denotes the number density of *pure* s at the specified temperature and pressure.

Having defined the Gibbs energy of solvation in equation (1.10), we can proceed to define all the other thermodynamic quantities of solvation using standard thermodynamic relationships. The most important quantities are the first derivatives of the free energy, i.e.,

$$\Delta S_s^* = - (\partial \Delta G_s^* / \partial T)_P \tag{1.17}$$

$$\Delta H_s^* = \Delta G_s^* + T\Delta S_s^* \tag{1.18}$$

$$\Delta V_s^* = (\partial \Delta G_s^* / \partial P)_T \tag{1.19}$$

One may also add the Helmholtz energy of solvation ($\Delta A_s^* = \Delta G_s^* - P\Delta V_s^*$) and the internal energy of solvation ($\Delta E_s^* = \Delta H_s^* - P\Delta V_s^*$). However, we shall see in Section 3.21 that for most of the specific systems discussed in this book the term $P\Delta V_s^*$ is usually negligible with respect to ΔH_s^* or ΔG_s^*. Therefore, the distinction between ΔG_s^* and ΔA_s^* or between ΔH_s^* and ΔE_s^* is usually insignificant.

In principle, second- and higher-order derivatives of the solvation Gibbs energy may also be defined by standard methods. These are potentially important and informative quantities. However, owing to lack of sufficient experimental data, it is not possible at present to evaluate these quantities. A brief survey of some formal definitions of these quantities in terms of measurable quantities is presented in Section 1.9.

Finally, we note that all the thermodynamic quantities of solvation as defined above pertain to exactly the same process, the solvation process as defined in this section. We stress, however, that the process of solvation is *not* an experimentally feasible one, i.e., we cannot carry out this process in the laboratory. This is probably the main reason why it has evaded the attention of so many researchers in the field of solution chemistry.

Being a molecular process, the solvation process cannot be handled within the realm of classical thermodynamics. It is accessible only through statistical mechanical considerations. Fortunately, as we shall see in the following section, statistical mechanics does provide a simple connection between solvation quantities and experimentally measurable quantities.

On the other hand, the fact that the solvation process, as defined above, is a molecular process makes it a powerful tool to probe both structural and energetic aspects of the local environment of a solvaton in various condensed phases. This will be amply demonstrated by examples in Chapter 2.

1.3. CALCULATION OF THE THERMODYNAMIC QUANTITIES OF SOLVATION FROM EXPERIMENTAL DATA

Having defined the solvation process in Section 1.2, we now turn to the question of evaluating the pertinent thermodynamic quantities from experimental data.[†] We have stressed in Section 1.2 that the solvation process is not experimentally feasible simply because we cannot fix the position of a molecule in a given system. Fortunately, we can carry out the solvation process in a thought experiment. Furthermore, using some simple results from statistical mechanics, we can express all the relevant thermodynamic quantities of solvation in terms of measurable quantities.

We discuss in this section the case of a solvaton s which does not undergo dissociation in any of the phases g and l. The more complicated case of dissociable solvatons (such as ionic solutes) will be discussed separately in Section 1.6.

Consider two phases α and β in which s molecules are distributed. We do not impose any restrictions on the concentration of s in the two phases. The only assumption being made is the applicability of classical statistical mechanics. At equilibrium, assuming that the two phases are at the same temperature and pressure, we have the following equation for the chemical potential of s in the two phases:

$$\mu_s^\alpha = \mu_s^\beta \tag{1.20}$$

In the traditional thermodynamic treatment one usually imposes the restriction of a very dilute solution of s in the two phases. However, we shall proceed by using the very general expression (1.6) for the chemical potential of s in the two phases. Applying equation (1.6) to equation (1.20), we obtain

$$\mu_s^{*\alpha} + kT \ln \rho_s^\alpha = \mu_s^{*\beta} + kT \ln \rho_s^\beta \tag{1.21}$$

[†] This procedure should be distinguished from a theoretical calculation based only on molecular properties of the molecules involved.

or, after rearrangement,

$$\Delta G_s^{*\beta} - \Delta G_s^{*\alpha} = (\mu_s^{*\beta} - \mu_s^{*ig}) - (\mu_s^{*\alpha} - \mu_s^{*ig})$$

$$= \mu_s^{*\beta} - \mu_s^{*\alpha} = kT\ln(\rho_s^{\alpha}/\rho_s^{\beta})_{eq} \qquad (1.22)$$

Here ρ_s^{α} and ρ_s^{β} are the number densities of s in the two phases, respectively, at equilibrium. Relation (1.22) provides a very simple way of computing the *difference* in the solvation Gibbs energies of s in the two phases α and β from the measurement of the two densities ρ_s^{α} and ρ_s^{β} at equilibrium.

It is noteworthy that in equation (1.22) the momentum partition function Λ_s^3 has been cancelled out. (This follows from the assumption of classical statistical mechanics.) Hence in equation (1.22), there seems to be no trace of molecular properties. This fact might lead to confusing relation (1.22) with a similar (but not identical) relation based on thermodynamic considerations only [see Section 1.4, in particular relations (1.42) and (1.25)].

In Section 1.6 we shall encounter a generalization of relation (1.22) for dissociable molecules, in which some of the molecular properties of the molecules must be used. The most useful particular example of (1.22) is when one of the phases, say α, is ideal gas. In such a case $\Delta G_s^{*\alpha} = 0$ and relation (1.22) reduces to

$$\Delta G_s^{*\beta} = kT\ln(\rho_s^{ig}/\rho_s^{\beta})_{eq} \qquad (1.23)$$

We note, however, that there is no restriction on the density of s in the phase β, but ρ_s^{ig} must be low enough to ensure that this phase (α) is an ideal gas. In many of the examples worked out in Chapter 2, we shall assume the ideality of the gaseous phase. A specific example is a liquid–vapor equilibrium in a one-component system. If the vapor pressure is low enough, we may safely assume that the vapor behaves as an ideal gas. In such a case we rewrite equation (1.23) as

$$\Delta G_s^{*p} = kT\ln(\rho_s^{ig}/\rho_s^{p})_{eq} \qquad (1.24)$$

where ΔG_s^{*p} is the solvation Gibbs energy of s in its own pure liquid [see equation (1.16)]. Examples of application of these relations are noble gases (Section 2.4) or pure liquid water (Section 2.11).

Another extreme case is the very dilute solution of s in the phase β, say argon in water, for which we have the limiting form of equation (1.23) that reads

$$\Delta G_s^{*\circ} = kT\ln(\rho_s^{ig}/\rho_s^{\beta})_{eq} \qquad (1.25)$$

This relation is identical in *form* to an equation derived from thermodynamics. However, it must be stressed that *conceptually* the two relations differ from each other, and further elaboration on this point is deferred to Section 1.4.

Once we have calculated the Gibbs energy of solvation through one of the relations cited above, it is a straightforward matter to calculate other thermodynamic quantities of solvation using the standard relations [(1.17) through (1.19)]. For instance, the solvation entropy might be calculated from the temperature dependence of the $\Delta G_s^{*\beta}$, or more specifically

$$\Delta S_s^{*\beta} = - \left\{ \frac{\partial}{\partial T} \left[kT \ln(\rho_s^{ig}/\rho_s^{\beta})_{eq} \right] \right\}_P \qquad (1.26)$$

We note that the differentiation in this equation is carried out at constant pressure. One must distinguish between this derivative and the derivative along the liquid–vapor equilibrium line. The relation between the two quantities is discussed in Section 1.5.

In this section we have derived the basic relations between the solvation thermodynamic quantities and experimental data (here, densities only). However, there exists a great deal of thermodynamic data tabulated under the heading of (conventional) solvation thermodynamics. We have taken advantage of the availability of these tables to convert from conventional to solvation quantities. The conversion procedure is outlined in Sections 1.4 and 1.5.

1.4. COMPARISON BETWEEN SOLVATION THERMODYNAMICS AND CONVENTIONAL STANDARD THERMODYNAMICS OF SOLUTION

In this section we present a detailed comparison between our *solvation* quantities on the one hand, and the conventional standard thermodynamics of solutions on the other. The latter are quite often also referred to as solvation quantities. As we shall demonstrate in this section, the conventional quantities are in principle inadequate measures of solvation quantities.

There are quite a few conventional quantities that have been employed in the literature. We shall discuss in this section only three of these, which we believe to be the most frequently used. Some general comments at the end of this section apply also to some less frequently used quantities.

In order to avoid any (understandable) confusion, a special notation will be used for the various conventional processes (e.g., x-process, ρ-

process, etc.). The superscript asterisk is reserved for the solvation process as defined in Section 1.2.

Let C_s be any concentration unit used to measure the concentration of s in the system. The most common units are the molarity ρ_s, the molality m_s, and the mole fraction x_s. (These are defined, along with interconversion relations, in Section 3.20. Note, also, that we shall often refer to ρ_s as either the molar or the number density. The two differ by the Avogadro number. It should be clear from the context to which of these we are referring in a particular case.)

In the conventional thermodynamic approach, we make use of the fact that Henry's law is obeyed in the limit of a very dilute solution of s. In this concentration range the CP of the species s (presuming it to be undissociable) is given by

$$\mu_s = \mu_s^{\circ c} + kT \ln C_s \qquad (1.27)$$

where $\mu_s^{\circ c}$ is referred to as the *standard* CP of s, based on the concentration scale C. This is formally defined as the limit

$$\mu_s^{\circ c} = \lim_{C_s \to 0} \ (\mu_s - kT \ln C_s) \qquad (1.28)$$

In all the conventional processes to be discussed below, it is important to bear in mind that the so-called (conventional) standard quantities apply only to very dilute solutions of s in the system. This is a very severe limitation when use is made of these quantities as measures of "solvation" of s.

From the molecular point of view, the validity of the limiting behavior of μ_s in equation (1.27) (Henry's law) is attained whenever the *solute* s is so dilute that one can safely neglect solute–solute correlations. (The latter is sometimes referred to, inadvisedly, as the absence of solute–solute interaction.)

Let α and β be two phases in which the concentrations of s are C_s^α and C_s^β, respectively. If the limiting expression (1.27) applies for both phases, we may define the Gibbs energy change for the process of transferring *one* s from α to β as

$$\Delta G \begin{bmatrix} \alpha \to \beta \\ C_s^\alpha, C_s^\beta \end{bmatrix} = \mu_s^\beta - \mu_s^\alpha = \mu_s^{\circ c\beta} - \mu_s^{\circ c\alpha} + kT \ln(C_s^\beta / C_s^\alpha) \qquad (1.29)$$

In all that follows we shall assume that the temperature and pressure are the same in the two phases. Hence, these will be omitted from our notation. On the left-hand side of relation (1.29), we do specify the concentrations of s in the two phases. These can be chosen freely as long as they are within the range of validity of equation (1.27).

Next, we make a *choice* of a standard process. In principle one can choose any specific values of C_s^α and C_s^β to characterize this standard process. A simple and very common choice is $C_s^\alpha = C_s^\beta$, for which relation (1.29) reduces to

$$\Delta G \begin{bmatrix} \alpha \to \beta \\ C_s^\alpha = C_s^\beta \end{bmatrix} = \mu_s^{\circ c\beta} - \mu_s^{\circ c\alpha} \tag{1.30}$$

This quantity is measurable provided we can measure the ratio C_s^β/C_s^α at equilibrium between the two phases α and β and provided both C_s^α and C_s^β at equilibrium are small enough to ensure ideal behavior. If these conditions are met, then in applying relation (1.29) to the case of equilibrium between the two phases, where $\mu_s^\alpha = \mu_s^\beta$, we obtain

$$\mu_s^{\circ c\beta} - \mu_s^{\circ c\alpha} = - kT \ln(C_s^\beta/C_s^\alpha)_{eq} \tag{1.31}$$

Thus equation (1.31) provides the means of measuring the standard Gibbs energy of transfer of s from α to β at $C_s^\alpha = C_s^\beta$.

We now define three special cases which are most commonly employed in the literature.

(1) The *ρ-process*. This is the process of transferring one s molecule from an ideal-gas phase into an ideal solution (Henry's law) at fixed temperature and pressure and such that $\rho_s^l = \rho_s^g$. This is essentially a special case of the general process described in relation (1.30) with the choice of the number (or molar) concentration units for C_s.

For this case we have

$$\Delta G_s(\rho\text{-process}) = \mu_s^{\circ\rho l} - \mu_s^{\circ\rho g} \tag{1.32}$$

where $\Delta G_s(\rho\text{-process})$ is a shorthand notation for

$$\Delta G \begin{pmatrix} g \to l \\ \rho_s^g = \rho_s^l \end{pmatrix}$$

The superscript "$\circ\rho l$" stands for *standard*, *ρ-units*, and *liquid phase*.

The relation between $\Delta G_s(\rho\text{-process})$ and the solvation Gibbs energy ΔG_s^* may be obtained by applying relation (1.6) to the two phases, resulting in

$$\Delta G_s(\rho\text{-process}) = \mu_s^{*l} - \mu_s^{*ig} \tag{1.33}$$

Comparison of expressions (1.33) and (1.10) yields the equality

$$\Delta G_s(\rho\text{-process}) = \Delta G_s^* \tag{1.34}$$

This is a noteworthy relation. It states that the Gibbs energy of the ρ-process and of the solvation process are equal in magnitude. However, one should carefully note that the two quantities appearing in equation (1.34) correspond to two different processes. Furthermore, this equality holds only for the Gibbs energies of the two processes. As we shall see below, such an equality does not exist for other thermodynamic quantities that correspond to these two processes. Second, the quantity $\Delta G_s(\rho\text{-process})$ may be determined by equation (1.31) (with $C = \rho$) only for the limiting case of very dilute solutions. The quantity ΔG_s^*, on the other hand, is obtainable from equation (1.23) for *any* concentration of s in the two phases.

(2) The *x-process*. This is defined as the process of transferring an s molecule from an ideal gas at 1 atm pressure to a hypothetical dilute-ideal solution in which the mole fraction of s is unity. (The temperature T and pressure 1 atm are the same in the two phases.)

The relation between the standard Gibbs energy of the x-process and the solvation Gibbs energy is obtained from the general expression (1.6), i.e.,

$$\Delta G_s(x\text{-process}) = [\mu_s^{*l} - \mu_s^{*g} + kT\ln(\rho_s^l/\rho_s^g)]_{\substack{P_s = 1 \\ x_s = 1}} \qquad (1.35)$$

where we must substitute $P_s = 1$ atm and $x_s = 1$ in equation (1.35). To achieve that we transform variables as follows. Since the x-process applies for ideal gases, we put

$$\rho_s^g = P_s/kT \qquad (1.36)$$

where P_s is the partial pressure of s in the gaseous phase. Furthermore, the x-process applies to dilute-ideal solutions; hence we may put

$$x_s^l = \frac{\rho_s^l}{\Sigma\rho_i^l} \approx \frac{\rho_s^l}{\rho_B^l} \qquad (1.37)$$

where ρ_B^l is the number density of the solvent B.

With these transformations of the variables we rewrite equation (1.35) in the form

$$\begin{aligned}
\Delta G_s(x\text{-process}) &= [\mu_s^{*l} - \mu_s^{*g} + kT\ln(x_s^l \rho_B^l kT/P_s)]_{\substack{P_s = 1 \\ x_s = 1}} \\
&= \mu_s^{*l} - \mu_s^{*g} + kT\ln(kT\rho_B^l) \\
&= \Delta G_s^* + kT\ln(kT\rho_B^l) \qquad (1.38)
\end{aligned}$$

Note that since we put $P_s = 1$ atm, $kT\rho_B^l$ must also be expressed in units of atmospheres.

Equation (1.38) is the required connection between the Gibbs energy of the x-process and the solvation Gibbs energy. Again, we note that this equality holds only for the ideal gas and ideal solutions.

(3) The *m-process*. This is defined as the process of transferring one s molecule from an ideal-gas phase at 1 atm pressure to a hypothetical ideal solution in which the *molality* of s is unity. (The temperature T and the pressure of 1 atm are the same in the two phases.)

The basic connection between the Gibbs energies of the m-process and the solvation process is again obtained from equation (1.6) as follows:

$$\Delta G_s(m\text{-process}) = [\mu_s^{*l} - \mu_s^{*g} + kT \ln(\rho_s^l/\rho_s^g)]_{\substack{P_s=1 \\ m_s=1}} \qquad (1.39)$$

Assuming that the gas phase is ideal, we can use the transformation (1.36). Furthermore, for very dilute solutions of s and assuming for simplicity that the solvent B is a one-component liquid with density ρ_B^l and molecular mass M_B, we can write the transformation from ρ_s^l into molality units m_s as follows:

$$\rho_s^l = M_B \rho_B^l m_s/1000 \qquad (1.40)$$

Substitution of relations (1.36) and (1.40) into equation (1.39) yields

$$\begin{aligned}\Delta G_s(m\text{-process}) &= [\mu_s^{*l} - \mu_s^{*g} + kT \ln(M_B \rho_B^l m_s kT/P_s \cdot 1000)]_{\substack{P_s=1 \\ m_s=1}} \\ &= \Delta G_s^* + kT \ln(M_B \rho_B^l kT/1000)\end{aligned} \qquad (1.41)$$

which is the required relation between the Gibbs energy changes for the m-process and for the solvation process.

For convenience, we now summarize the three relationships between the various standard Gibbs energies of transfer and the Gibbs energy of solvation:

$$\Delta G_s(\rho\text{-process}) = \Delta G_s^* \qquad (1.42)$$

$$\Delta G_s(x\text{-process}) = \Delta G_s^* + kT \ln(kT\rho_B^l) \qquad (1.43)$$

$$\Delta G_s(m\text{-process}) = \Delta G_s^* + kT \ln(M_B \rho_B^l kT/1000) \qquad (1.44)$$

We note that since all the three processes defined above refer to ideal behavior of both phases, ΔG_s^* in equations (1.42) to (1.44) are essentially equal to $\Delta G_s^{*\circ}$ as defined in equation (1.15).

Some general comments are now in order. First, we observe that relation (1.42) is relatively the simplest. We might be tempted to conclude that a study of the ρ-process is equivalent to the study of the solvation process. Such a conclusion, though valid for the Gibbs energies of the two processes, is not valid for any other thermodynamic quantity, as we shall see below. This apparent equivalency has caused some confusion in the literature.

As is evident from these relations, the p-process is the closest to the solvation process. Furthermore, from the thermodynamic point of view, it is the simplest of the three since it does not involve the so-called hypothetical states. (The latter are quite awkward and a critical examination of them may be found elsewhere.[1,4])

In relations (1.43) and (1.44) we see that the difference between the Gibbs energy of the thermodynamic process (either x-process or m-process) and the solvation process is a constant quantity depending on the properties of the solvent. This means that if we wish to compare various *solutes* in the same solvent B, we may disregard the constant quantities on the rhs of equations (1.43) and (1.44). But what if we wish to study a single solute in various solvents? Here we encounter a serious problem because of the special way the solvent density features in the quantities on the rhs of equations (1.43) and (1.44). To demonstrate the difficulty, suppose we wish to study the solvation Gibbs energy of a given solute s in a series of solvents B having decreasing densities ρ_B. One can easily show that ΔG_s^* will tend to zero as $\rho_B^l \to 0$. This is clearly the behavior we should expect from a quantity that measures the extent of the Gibbs energy of interaction between s and its environment. In the extreme case when $\rho_B^l = 0$, we have $\Delta G_s^* = 0$ as it should be [this actually follows from the definition of ΔG_s^* in (1.10), but see also Section 3.10]. This behavior holds true for all other quantities of solvation (see below). On the other hand, a glance at equations (1.43) and (1.44) shows that when $\rho_B^l \to 0$ both $\Delta G_s(x$-process$)$ and $\Delta G_s(m$-process$)$ will tend to minus infinity. This behavior is clearly unacceptable for a quantity that is presumed to measure the Gibbs energy of interaction between s and its environment. We shall see below that such divergent behavior is exhibited by other thermodynamic quantities corresponding to the x-process and the m-process. It is for this reason that both of these quantities cannot in principle serve as *bona fide* measures of the solvation Gibbs energy.

Next, we derive the relations between some of the other thermodynamic quantities associated with the processes defined above.

(1) *Entropy*. As before, our starting point is the general equation (1.6) for the CP. The partial molar (or molecular) entropy of s is obtained from

$$\bar{S}_s = -\partial\mu_s/\partial T = -\partial\mu_s^*/\partial T - k\ln\rho_s\Lambda_s^3 - kT\partial\ln(\rho_s\Lambda_s^3)/\partial T \qquad (1.45)$$

Note that the differentiation is taken at constant pressure and composition of the system.

We denote by S_s^* the entropy change that corresponds to the process of adding one s molecule to a fixed position in the system. On performing the differentiation with respect to temperature in equation (1.45), we obtain

$$\bar{S}_s = S_s^* - k \ln \rho_s \Lambda_s^3 + kT\alpha_P + \tfrac{3}{2}k \qquad (1.46)$$

where $\alpha_P = V^{-1}\partial V/\partial T$ is the thermal expansion coefficient of the system at constant pressure.

Applying equation (1.46) for an ideal-gas phase and for an ideal-dilute solution, we may derive the entropy changes associated with the standard processes as defined above. For the solvation process, we simply have

$$\Delta S_s^* = S_s^{*l} - S_s^{*g} = -\partial \Delta G_s^*/\partial T \qquad (1.47)$$

(In most cases the superscript g for "gas" is understood to stand for ideal gas. If the gaseous phase is not ideal, then ΔS_s^* is the difference in the solvation entropy between the two phases.)

For the ρ-process we have

$$\Delta S_s(\rho\text{-process}) = [\bar{S}_s^l - \bar{S}_s^g]_{\rho_s^l = \rho_s^g} = S_s^{*l} - S_s^{*g} + kT\alpha_P^l - kT\alpha_P^g$$
$$= \Delta S_s^* + kT\alpha_P^l - k \qquad (1.48)$$

where we put $\alpha_P^g = T^{-1}$ for the ideal-gas phase.

Similarly, for the x-process and the m-process we have, respectively,

$$\Delta S_s(x\text{-process}) = [\bar{S}_s^l - \bar{S}_s^g]_{\substack{P_s = 1 \\ x_s = 1}} = S_s^{*l} - S_s^{*g} - k\ln(kT\rho_B^l) + kT\alpha_P^l - k$$
$$= \Delta S_s^* - k\ln(kT\rho_B^l) + kT\alpha_P^l - k \qquad (1.49)$$

and

$$\Delta S_s(m\text{-process}) = [\bar{S}_s^l - \bar{S}_s^g]_{\substack{P_s = 1 \\ m_s = 1}} = \Delta S_s^* - k\ln(M_B\rho_B^l kT/1000) + kT\alpha_P^l - k \qquad (1.50)$$

In contrast to the case of the Gibbs energies, where we encountered only *three* different quantities that correspond to *four* distinctly different processes [equations (1.42) through (1.44)], here we have *four* distinctly different quantities. The entropy change for the ρ-process is *unequal* to the entropy of solvation. We note also that $\Delta S_s(\rho\text{-process})$ cannot be obtained by direct differentiation of $\Delta G_s(\rho\text{-process})$ with respect to the temperature. The reason is that equality (1.42) holds only at the point for which $\rho_s^l = \rho_s^g$. Changing the temperature will, in general, change the densities of the two phases by different amounts. This leads to the additional term $kT\alpha_P^l - k$ in equation (1.48).

A glance at the expressions for $\Delta S_s(x\text{-process})$ and $\Delta S_s(m\text{-process})$ shows that both contain the solvent density ρ_B^l under the logarithm sign. Thus for a series of solvents with decreasing densities, both $\Delta S_s(x\text{-process})$ and $\Delta S_s(m\text{-process})$ will diverge to infinity, clearly an undesirable feature for a quantity that is presumed to measure the solvation entropy of a molecule s. On the other hand, ΔS_s^* tends to zero as the solvent density decreases to zero. In addition to this unacceptable behavior of $\Delta S_s(x\text{-process})$ and $\Delta S_s(m\text{-process})$, all of these standard entropies of transfer contain the term $kT\alpha_P^l - k$, which is irrelevant to the solvation process of the molecule s.

(2) *Enthalpy.* To obtain the enthalpies of the various processes, we simply form the combination $\bar{H}_s = \mu_s + T\bar{S}_s$ and apply it to the four processes of interest. The results are

$$\Delta H_s^* = H_s^{*l} - H_s^{*g} = \Delta G_s^* + T\Delta S_s^* \tag{1.51}$$

$$\Delta H_s(p\text{-process}) = \Delta H_s(x\text{-process}) = \Delta H_s(m\text{-process})$$
$$= \Delta H_s^* + kT^2\alpha_P^l - kT \tag{1.52}$$

Here we see the well-known property that the enthalpy changes for the three standard processes are identical. This follows from the assumption of ideality introduced in the definition of these processes. On the other hand, these three processes produce an enthalpy change which *differs* from the solvation enthalpy by the quantity $kT^2\alpha_P^l - kT$. Again, although this quantity is not divergent at $\rho_B^l \to 0$, it is certainly irrelevant to the process of solvation of a molecule s.

(3) *Volume.* Taking the derivative of the general expression for the chemical potential [e.g., relation (1.6)] with respect to pressure, we obtain the partial molar (or molecular) volume of s, i.e.,

$$\bar{V}_s = (\partial\mu_s/\partial P)_T = (\partial\mu_s^*/\partial P)_T + kT(\partial \ln \rho_s/\partial P)_T$$

$$= V_s^* + kT\beta_T^l \tag{1.53}$$

where V_s^* is the volume change due to the addition of one s molecule at a fixed position in the system, and β_T^l is the isothermal compressibility of the phase l,

$$\beta_T^l = -\frac{1}{V}\left(\frac{\partial V}{\partial P}\right)_T \tag{1.54}$$

which, for an ideal-gas phase, reduces to

$$\beta_T^{ig} = 1/P \tag{1.55}$$

The volume changes for the four processes of interest are

$$\Delta V_s^* = \partial \Delta G_s^* / \partial P = V_s^{*l} - V_s^{*g} \tag{1.56}$$

$$\Delta V_s(\rho\text{-process}) = \Delta V_s^* + [kT(\beta_T^l - P^{-1})]_{\rho^l = \rho^g} \tag{1.57}$$

$$\Delta V_s(x\text{-process}) = \Delta V_s(m\text{-process}) = \Delta V_s^* + kT(\beta_T^l - 1/\text{atm}) \tag{1.58}$$

Here the volume changes for the last two processes are identical. We note also that for the liquid phases at room temperaturs β_T^l is much smaller than 1 $(\text{atm})^{-1}$ (e.g., for water at 0 °C, $kT\beta_T^l \sim 1$ cm^3 mol^{-1}, $\Delta V_s^* \sim 20$ cm^3 mol^{-1}, and $kT/\text{atm} \sim 2 \times 10^4$ cm^3 mol^{-1}). Similarly, in equation (1.57) $\beta_T^l \ll P^{-1}$ (where we should take the limit of an ideal-gas phase). Thus, the volume change for the three standard processes is dominated by the terms which originate from the ideal-gas compressibility. Because of this feature, it is common to abandon these processes when studying the volume of solvation. Almost all researchers who study solvation phenomena apply one of these standard processes for quantities like the Gibbs energy, entropy, and enthalpy of solvation. However, for the volume of solvation they switch to partial molar volumes at infinite dilution. The latter clearly corresponds to a *different* process. Similar difficulties appear in the study of higher-order derivatives of the free energy.

The solvation process as defined in Section 1.2 can be applied uniformly to all the thermodynamic quantities of solvation. Obviously, this is a convenient feature of the solvation process which is not shared by other conventional standard processes.

From the quantities derived above, one may construct the internal energy of solvation ($\Delta E_s^* = \Delta H_s^* - P\Delta V_s^*$) and the Helmholtz energy of solvation ($\Delta A_s^* = \Delta G_s^* - P\Delta V_s^*$) and compare these with the corresponding conventional quantities. As noted in Section 1.2, the difference between ΔE_s^* and ΔH_s^* and between ΔA_s^* and ΔG_s^* is usually very small and may be neglected for most systems of interest discussed in this book (see also Section 3.21).

Of course, one may easily go beyond first-order derivatives of the Gibbs energy. One can define the compressibility, heat capacity, thermal expansion, and so on, for the process of solvation. These quantities are of potential interest in the study of solvation phenomena. However, for lack of sufficient experimental data we cannot compute any of these quantities at present. Some formal definitions of these quantities are given in Section 1.9.

We now summarize the essential differences between the present approach to the study of solvation phenomena and the conventional one based on various standard processes. First and foremost is the simple fact that the solvation process as defined in Section 1.2 is the most direct means of probing the interaction of a solvaton with its environment. The pertinent thermodynamic quantities of solvation tend to zero when the solvent density goes to zero (i.e., when there is no interaction between the solvaton and its environment). This is not the case for the conventional thermodynamic quantities, some of which even diverge to plus or minus infinity in this limit. Furthermore, by adopting the solvation process we achieve both a generalization and a uniformity in application of this concept. The generalization involves the extension of the range of concentration of s for which the solvation thermodynamics may be studied: from very dilute s up to pure liquid s. The uniformity involves the application of the *same* process for all thermodynamic quantities.

Finally, as a concluding remark we may add that, once we adopt the definition of the solvation process as given in Section 1.2, we can forget about all standard states. This is a drastic simplification compared to the existing situation where, in addition to specifying the thermodynamic variables of the system, one must also announce the standard states that have been chosen.[†] In our approach, it will be abundantly clear from the examples given in Chapter 2 that we may speak of "solvation of argon in water," "argon in benzene," or "argon in argon" (at a specified temperature and pressure). This will be sufficient to clarify which environment of the solvaton is being studied. No standard state or standard process need be invoked!

1.5. SOME FURTHER RELATIONSHIPS BETWEEN SOLVATION THERMODYNAMICS AND AVAILABLE THERMODYNAMIC DATA

In Section 1.3 we outlined the fundamental connections between solvation thermodynamics and experimental data. However, in many cases thermodynamic data are tabulated that may be transformed into solvation thermodynamics without going through the fundamental relationships. Some of these transformations are presented here. A separate treatment of ionic solutions is presented in Section 1.6.

(1) *Very dilute solutions of s in l.* These are the only systems for which the concept of solvation in the conventional meaning has been applied. For these systems there are many publications of tables of thermodynamics of solution (or solvation) which pertain to one of the

---

[†] Unfortunately, even this essential requirement is not fulfilled in many research articles where tables of, say, entropy of solvation are given without specifying to which process they refer.

processes discussed in Section 1.4. For these cases all the conversion formulas have been derived in the previous section. These were actually used to compute some of the quantities presented in Chapter 2. Here we add one more connection with a very commonly used quantity, the Henry's law constant. In its most common form it is defined by

$$K_H = \lim_{x_s^l \to 0} (P_s/x_s^l) \tag{1.59}$$

where P_s is the partial pressure, x_s^l is the mole fraction of s in the system (if definable), and the limit takes x_s^l into the range where Henry's law becomes valid.

The general expression for the solvation Gibbs energy in this case is [see equation (1.25)]

$$\Delta G_s^{*\circ} = kT \ln(\rho_s^{ig}/\rho_s^l)_{eq} \tag{1.60}$$

Assuming that we have a sufficiently dilute solution of s in l, such that Henry's law in the form $P_s = K_H x_s^l$ is obeyed, we can transform equation (1.60) into

$$\Delta G_s^{*\circ} = kT \ln(P_s/kTx_s^l\rho_B^l)$$
$$= kT \ln(K_H/kT\rho_B^l) \tag{1.61}$$

This is the required connection between the tabulated values of K_H and the solvation Gibbs energy. (Here we have assumed, for simplicity, that the solvent consists of one component with a number density ρ_B^l. If the solvent contains several components, excluding s, then ρ_B^l should be replaced by the sum of the number densities of all the other components.) We note also that from relations (1.43) and (1.61) we also have

$$\Delta G_s(x\text{-process}) = kT \ln K_H \tag{1.62}$$

Thus, information on K_H is essentially equivalent to information on the Gibbs energy change for the x-process.

(2) *Concentrated solutions.* For solutions (or rather mixtures) of s at higher concentrations beyond the realms of Henry's law, we depart from the traditional notion of solvation and must use the definition as presented in Section 1.2.

There exists a variety of data that measure the extent of deviation from dilute ideal solutions. These include tables of activity coefficients, osmotic coefficients, and excess functions. All of these may be used to compute solvation thermodynamic quantities (as in fact has been done for some specific systems discussed in Chapter 2).

Our starting point is the general expression for the CP of s in the liquid phase l, e.g., equation (1.6):

$$\mu_s = \mu_s^{*l} + kT \ln \rho_s^l \Lambda_s^3 \tag{1.63}$$

The same CP in the same system is expressed in conventional thermodynamics as

$$\mu_s = \mu_s^{\circ\rho} + kT \ln \rho_s^l + kT \ln \gamma_s^\rho \tag{1.64}$$

where γ_s^ρ is the activity coefficient that measures deviations with respect to the ideal-dilute behavior, based on the number density ρ_s as concentration units;[†] $\mu_s^{\circ\rho}$ is the conventional standard CP of s in the ρ concentration scale and is formally defined by

$$\mu_s^{\circ\rho} = \lim_{\rho_s^l \to 0} (\mu_s - kT \ln \rho_s^l) \tag{1.65}$$

Substitution of equation (1.63) into (1.65) yields

$$\mu_s^{\circ\rho} = \lim_{\rho_s^l \to 0} (\mu_s^{*l} + kT \ln \rho_s^l \Lambda_s^3 - kT \ln \rho_s^l)$$

$$= \mu_s^{*\circ l} + kT \ln \Lambda_s^3 \tag{1.66}$$

which is the connection between the conventional standard CP $\mu_s^{\circ\rho}$ and the PCP of s at infinite dilution $\mu_s^{*\circ l}$. By using relations (1.63), (1.64), and (1.66), we arrive at the final expression:

$$\mu_s^{*l} - \mu_s^{*\circ l} = kT \ln \gamma_s^\rho \tag{1.67}$$

This quantity is equivalent to the difference between the solvation Gibbs energy of s in the phase l and the solvation Gibbs energy of s in the same phase except for taking the limit $\rho_s \to 0$. In the notation of Section 1.2 we may rewrite this quantity as

$$\Delta G_s^* - \Delta G_s^{*\circ} = \mu_s^{*l} - \mu_s^{*\circ l} = kT \ln \gamma_s^\rho \tag{1.68}$$

Thus, from the activity coefficient (based on the ρ-concentration scale), one can compute the solvation Gibbs energy of s in a liquid phase l (containing any quantity of s), relative to the solvation Gibbs energy of s

---

[†] Care must be exercised to clearly distinguish between several different activity coefficients. This point is elaborated in Chapter 4 of Ben-Naim.[4]

in the same phase but at $\rho_s \to 0$. This quantity is quite useful whenever the infinite dilute limit is already known.

Relation (1.68) holds only for the activity coefficient as defined in equation (1.64), i.e., based on the number density scale. However, it is quite a simple matter to use any other activity coefficient to extract the same information. Let C_s be any other concentration units (e.g., molality, mole fraction, etc.). We write the general conversion relation between C_s and ρ_s as

$$\rho_s = T_s^c C_s \qquad (1.69)$$

Some specific forms of T_s^c are given in Section 3.20. The general expression of the CP (1.63) may be expressed in the C scale as

$$\mu_s = \mu_s^{*l} + kT \ln T_s^c C_s \Lambda_s^3 \qquad (1.70)$$

in the limit of a very dilute solution $\rho_s \to 0$, we write

$$\mu_s = \mu_s^{*\circ l} + kT \ln T_s^{c,0} C_s \Lambda_s^3 \qquad (1.71)$$

where

$$T_s^{c,0} = \lim_{\rho_s \to 0} T_s^c \qquad (1.72)$$

The standard CP in the C scale is given by

$$\begin{aligned}
\mu_s^{\circ c} &= \lim_{\rho_s \to 0} (\mu_s - kT \ln C_s) \\
&= \lim_{\rho_s \to 0} (\mu_s^{\circ\rho} + kT \ln \rho_s \gamma_s^\rho - kT \ln C_s) \\
&= \mu_s^{\circ\rho} + kT \ln T_s^{c,0} \qquad (1.73)
\end{aligned}$$

Hence the relation between the two activity coefficients is

$$\begin{aligned}
kT \ln \gamma_s^c &= \mu_s - \mu_s^{\circ c} - kT \ln C_s = \mu_s - \mu_s^{\circ\rho} - kT \ln(T_s^{c,0} \rho_s / T_s^c) \\
&= kT \ln \gamma_s^\rho + kT \ln(T_s^c / T_s^{c,0}) \qquad (1.74)
\end{aligned}$$

and the corresponding connection with the solvation Gibbs energy is

$$\Delta G_s^* - \Delta G_s^{*\circ} = kT \ln \gamma_s^c + kT \ln(T_s^c / T_s^{c,0}) \qquad (1.75)$$

which may be used when activity coefficients based on any concentration scale are available.

The second important source of data available for multicomponent mixtures (say argon and krypton, or water and ethanol) is the so-called excess thermodynamic functions. These are equivalent to activity coefficients that measure deviations from symmetrical ideal solutions and should be distinguished carefully from activity coefficients that measure deviations from ideal-dilute solutions.[†]

In a symmetrical ideal (SI) solution the CP of each component is related to its mole fraction through a relation of the form

$$\mu_i = \mu_i^p + kT \ln x_i \tag{1.76}$$

where μ_i^p is the CP of the pure component i (i.e., the case $x_i = 1$ at the same temperature and pressure).

The SI behavior is realized by a variety of two-component systems of two "similar" species (the exact necessary and sufficient conditions for such behavior is discussed elsewhere[†]). Deviations from this behavior may be expressed by introducing either an activity coefficient γ_s^{SI} or an excess function; these are defined as

$$\mu_i = \mu_i^p + kT \ln x_i + kT \ln \gamma_i^{SI} = \mu_i^p + kT \ln x_i + \mu_i^E \tag{1.77}$$

From now on we shall use only the excess CP as defined by relation (1.77). The total excess Gibbs energy of the entire system is given by

$$G^E = \sum_i N_i \mu_i^E \tag{1.78}$$

and the excess Gibbs energy per molecule of the mixture is defined by

$$g^E = G^E \Big/ \sum_i N_i = \sum_i x_i \mu_i^E \tag{1.79}$$

For some two-component systems, the quantity g^E is available as an analytical function of the composition of the system. Let A and B be the two components and x_A the mole fraction of A in the system:

$$x_A = N_A/(N_A + N_B) \tag{1.80}$$

When g^E is given as a function of x_A in the entire range of compo-

---

[†] A detailed elaboration on this matter may be found in Chapter 4 of Ben-Naim.[(4)]

sitions, one can recover both μ_A^E and μ_B^E using the following well-known procedure:[5]

$$\mu_A^E = (\partial G^E/\partial N_A)_{P,T,N_B} = \frac{\partial}{\partial N_A}[(N_A + N_B)g^E]$$

$$= (N_A + N_B)\partial g^E/\partial N_A + g^E \tag{1.81}$$

Transforming variables from N_A into x_A

$$\frac{\partial}{\partial N_A} = \frac{N_B}{(N_A + N_B)^2}\frac{\partial}{\partial x_A} \tag{1.82}$$

yields

$$\mu_A^E = g^E + x_B\partial g^E/\partial x_A \tag{1.83}$$

and similarly

$$\mu_B^E = g^E + x_A\partial g^E/\partial x_B \tag{1.84}$$

The connection between excess CPs and solvation Gibbs energies may be obtained as follows. We write both μ_A and μ_A^p using the general expression (1.6) and the thermodynamic expression (1.77) to obtain

$$\mu_A = \mu_A^p + kT\ln x_A + \mu_A^E = \mu_A^{*p} + kT\ln \rho_A^p\Lambda_A^3 x_A + \mu_A^E \tag{1.85}$$

$$\mu_A = \mu_A^* + kT\ln \rho_A\Lambda_A^3 \tag{1.86}$$

By comparing equation (1.85) with (1.86) we arrive at

$$\Delta G_A^* - \Delta G_A^{*p} = \mu_A^* - \mu_A^{*p} = kT\ln(\rho_A^p x_A/\rho_A) + \mu_A^E \tag{1.87}$$

Here we have the solvation Gibbs energy of A in the mixture, relative to the solvation of A in pure A (at the same temperature and pressure). This may be computed for any composition from knowledge of the excess CP μ_A^E and the densities of A in the mixture and in the pure component, ρ_A and ρ_A^p, respectively. A similar expression may be written for the second component B.

Sometimes, the densities ρ_A are not available in the entire range of compositions. Instead, data on excess volume are available. This may be used as follows. The excess volume per molecule of the mixture is given by

$$v^E = \frac{V^E}{N_A + N_B} = \frac{V - N_A V_A^p - N_B V_B^p}{N_A + N_B} = v_m - x_A V_A^p - x_B V_B^p \tag{1.88}$$

where V_A^p and V_B^p are the molar (or molecular) volumes of pure A and B, respectively, and v_m is the volume per molecule of the mixture given by $v_m = V/(N_A + N_B)$. Hence

$$\frac{\rho_A^p x_A}{n_{\scriptscriptstyle A}} = \frac{\rho_A^p}{\rho_A + \rho_B} = \frac{v_m}{V_A^p} = \frac{v^E + x_A V_A^p + x_B V_B^p}{V_A^p} \tag{1.89}$$

This may be used in equation (1.87) to calculate the relative solvation Gibbs energy of A. This procedure is employed for the computations presented in Sections 2.5 and 2.14.

(3) *Pure liquids*. The extreme limit of high density of s is the pure liquid. Of course, by changing the temperature or the pressure one changes the density of the liquid as well. However, normally the liquids we shall be dealing with are either at room temperature and 1 atm pressure or along the liquid–vapor coexistence equilibrium line.

Let *l* and *g* be the liquid and the gaseous phases of a pure component s at equilibrium. The Gibbs energy of transferring s from a fixed position in *g* into a fixed position in *l* is equal to the difference in the solvation Gibbs energies of s in the two phases, i.e.,

$$\mu_s^{*l} - \mu_s^{*g} = \Delta G_s^{*l} - \Delta G_s^{*g} = kT \ln(\rho_s^g/\rho_s^l)_{eq} \tag{1.90}$$

Thus, in general, knowing the densities of s in the two phases at equilibrium gives us only the difference in the solvation Gibbs energies of s in the two phases. However, in many cases, especially near the triple point, the density of the gaseous phase is quite low, in which case we may assume that $\Delta G_s^{*g} \approx 0$ and therefore relation (1.90) reduces to

$$\Delta G_s^{*p} = kT \ln(\rho_s^{ig}/\rho_s^l)_{eq}$$
$$= kT \ln(P_s/kT\rho_s^l)_{eq} \tag{1.91}$$

where P_s is the vapor pressure of s at the temperature T.

When evaluating other thermodynamic quantities of solvation from data along the equilibrium line, one must be careful to distinguish between derivatives at constant pressure and derivatives along the equilibrium line. The connection between the two is

$$\left(\frac{d\Delta G_s^*}{dT}\right)_{eq} = \left(\frac{\partial \Delta G_s^*}{\partial T}\right)_P + \left(\frac{\partial \Delta G_s^*}{\partial P}\right)_T \left(\frac{dP}{dT}\right)_{eq} \tag{1.92}$$

Here we use straight derivatives to indicate differentiation along the equilibrium line. The two coefficients on the rhs of equation (1.92) are now identified as the solvation entropy and volume, respectively; thus

$$\left(\frac{d\Delta G_s^*}{dT}\right)_{eq} = -\Delta S_s^{*p} + \Delta V_s^{*p}\left(\frac{dP}{dT}\right)_{eq} \tag{1.93}$$

Usually, data are available to evaluate both of the straight derivatives in equation (1.93). This is not sufficient, however, to compute both ΔS_s^{*p} and ΔV_s^{*p}. Fortunately, ΔV_s^{*p} may be obtained directly from data on molar volume and compressibility of the pure liquid. For the pure system s we have

$$\mu_s^p = \mu_s^{*p} + kT\ln\rho_s^p\Lambda_s^3 \tag{1.94}$$

Differentiation with respect to P at constant T yields

$$\bar{V}_s^p = \left(\frac{\partial\mu_s^p}{\partial P}\right)_T = V_s^{*p} + kT\beta_T^p \tag{1.95}$$

where β_T^p is the isothermal compressibility of the pure system. Since $V_s^{*ig} = 0$, we obtain

$$\Delta V_s^{*p} = \bar{V}_s^p - kT\beta_T^p \tag{1.96}$$

from which we can compute ΔV_s^{*p}. This may be used in equation (1.93) (presuming that ΔV_s^{*p} is approximately insensitive to small changes in pressure) to compute ΔS_s^{*p}. This procedure is employed in Sections 2.4 and 2.11.

1.6. SOLVATION OF IONIC SOLUTES

We devote a special section to discuss ionic solvation (or, more generally, solvation of dissociable molecules) because the basic relation between solvation Gibbs energy and experimental quantities is somewhat more complicated than the one we have developed for undissociable molecules.

In the study of ionic solvation there are essentially two new problems that are absent in the case of nondissociable solutes. First, how does one define and measure the various solvation quantities for the entire, electrically neutral molecule? Second, how does one separate the contribution of the single ions to the total solvation quantities? This section deals only with the answer to the first problem. The second question is still left open for various approaches that have been suggested in the literature. Perhaps we should say, however, that once we adopt one of the

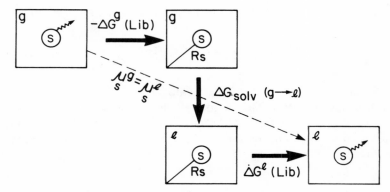

Figure 1.2. A schematic description of the process of solvation for a simple spherical solute s. The thick arrows correspond to the three steps as described in equation (1.97). The equilibrium is indicated by the dashed line.

suggested conventions for splitting into single ion contributions, the best place to implement these ideas would be within the framework of solvation thermodynamics as defined in this book (see also Sections 3.12 and 3.13).

By extending the definition given in Section 1.2, we define the *solvation process* of a molecule D (or dimer) that dissociates into two groups (or ions) A and B, as the process of transferring A and B from two fixed positions and infinite separation in the gaseous phase into two fixed positions and infinite separation in the liquid phase (T, P, and whatever compositions of the system are kept constant).

In this section we present only a qualitative description of the method we use to calculate the solvation Gibbs energy of the ionic solute. The full statistical mechanical derivation is deferred to Section 3.12. In order to appreciate the new difficulties that arise in the ionic case, we repeat the method used to relate the solvation Gibbs energy to experimental data, in a diagrammatic form as depicted in Figure 1.2.

We start with a solute s free to wander in an ideal-gas phase. We first freeze the translational degrees of freedom of s so that it is now constrained to a fixed position, say $\mathbf{R}_s$. Next, we transfer it from a fixed position in the gas to a fixed position in the liquid, and finally, we release the constraint on a fixed position. The Gibbs energy changes corresponding to these three steps (thick arrows in the diagram) are connected with the equilibrium condition $\mu_s^l - \mu_s^g = 0$ by the equation

$$0 = \mu_s^l - \mu_s^g = [-kT \ln \rho_s^g \Lambda_s^3] + [\Delta G_s^*] + [kT \ln \rho_s^l \Lambda_s^3] \qquad (1.97)$$

As we pointed out in Section 1.3, it is fortunate in this case that the momentum partition function Λ_s^3 cancels out, and hence leads to a relatively simple relation between ΔG_s^* and the ratio of densities [equation (1.23)].

Suppose now we have a dimeric molecule in the gaseous phase which is known to dissociate completely into two groups (or ions) A and B in the liquid phase (e.g., D may be HCl and A and B, H^+ and Cl^-, respectively). For this case we extend the diagram of Figure 1.2 to the one depicted in Figure 1.3. Here, four steps are invoved. First, we freeze the translational, rotational and vibrational degrees of freedom. The corresponding changes in Gibbs energy are $-kT \ln \rho_D^g \Lambda_D^3$, $kT \ln q_{rot}$ and $kT \ln q_{vib}$, respectively. Here, ρ_D^g is the density of the dimers in the gaseous phase, and Λ_D^3 is the momentum partition function of the dimer molecule; q_{rot} is the rotational partition function, and q_{vib} is the vibrational partition function, in which the energy levels are measured relative to the bottom of the potential energy (see also Section 3.12 for more details). After this step has been accomplished, we have the dimer D at a fixed position and orientation in the gaseous phase. In addition, the intramolecular separation between A and B is at the minimum of the potential-energy curve $R_{AB} = \sigma$.

Next, we separate the two groups from the distance $R_{AB} = \sigma$ to infinity, $R_{AB} = \infty$. This separation involves the potential energy $U_{AB}(\infty) - U_{AB}(\sigma)$. If A and B are ionic species, say H^+ and Cl^-, then we have to add the ionization energy to the cation and the electron affinity of the anion. From this stage we perform the solvation process as defined above, and finally, we release the constraint imposed on the fixed positions of A and B.

Combining the Gibbs energy changes involved in the four steps with the equilibrium condition $\mu_D^l - \mu_D^g = 0$, we arrive at the final relation:

$$0 = \mu_D^l - \mu_D^g = [-kT \ln \rho_D^g \Lambda_D^3 q_{rot}^{-1} q_{vib}^{-1}]$$

$$+ [U_{AB}(\infty) - U_{AB}(\sigma) + \Delta E(\text{ionization})]$$

$$+ [\Delta G_{A,B}^{*l}] + [kT \ln \rho_A^l \rho_B^l \Lambda_A^3 \Lambda_B^3] \qquad (1.98)$$

Here, each term in square brackets corresponds to the Gibbs energy change along one of the thick arrows, as indicated in Figure 1.3. The solvation Gibbs energy of the pair of ions is denoted by $\Delta G_{A,B}^{*l}$. If the ions do not have any internal degrees of freedom, then $\Delta G_{A,B}^{*l}$ is essentially the coupling work of A and B to the liquid phase l, i.e.,

$$\Delta G_{A,B}^{*l} = W(A, B \mid l) \qquad (1.99)$$

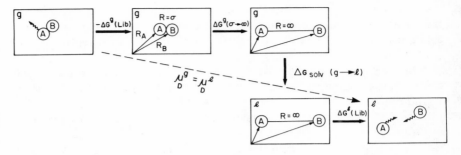

Figure 1.3. Schematic description of the solvation of a dissociable solute D from the gaseous phase into the liquid. The thick arrows corresponding to the four steps as described in equation (1.98). First, we "freeze" the translation, rotation, and vibration of the dimer D. We then separate the two ions to infinite distance. Next, we solvate the two ions, i.e., we transfer the two ions from fixed positions and infinite separation in the gas to fixed positions in the liquid. Finally, we release the two particles to wander in the liquid phase. The free energy change in each step corresponds to one term on the rhs of equation (1.98).

The quantity ΔE(ionization) combines the ionization energy and the electron affinity of the cation and the anion, respectively. For molecules that dissociate into neutral particles, we set ΔE(ionization) $= 0$.

In Section 2.17 we present some computations based on equation (1.98), from which we extract the solvation Gibbs energies $\Delta G^{*l}_{A,B}$. We have also used available tables of standard Gibbs energies of solution and Gibbs energies of formation to convert them into solvation quantities. As is obvious from equation (1.98), the calculation of $\Delta G^{*l}_{A,B}$ requires not only experimental data on the densities ρ^g_D, ρ^l_A, and ρ^l_B at equilibrium, but also some molecular quantities included in Λ^3_D, Λ^3_A, Λ^3_B, q_{rot}, and q_{vib}.

1.7. SOLVATION OF WATER IN DILUTE AQUEOUS SOLUTIONS

The thermodynamics of solvation of water in pure water is discussed in Section 2.11. This is a particular example of a solvation of a molecule s in its own liquid. However, liquid water holds a very special position among all the liquids that have been studied in physical chemistry. One particular aspect of liquid water is the relation between the structure of water on a molecular level and the anomalous properties of water as revealed to us in the laboratory on a macroscopic level. A closely related aspect, the effect of solutes on the structure of water, has been investigated by a multitude of experimental means.

Traditionally, the structural changes in water induced by a solute were investigated through the solvation properties of *solutes* in aqueous solutions. Never before has the solvation of a water molecule itself been studied, either in pure water or in aqueous solutions. This topic was simply inaccessible through the conventional concepts of solvation thermodynamics. However, with the definition of the solvation process as presented in Section 1.2, a new tool has become available to study the solvation properties of a water molecule both in pure water and in aqueous solutions.

In the present section we derive the appropriate tool by the use of which we can probe both structural and energetic effects of various solutes on the immediate surroundings of a water solvaton. We shall restrict ourselves here to the limiting effect of a solute on the solvation therodynamics of a water molecule (numerical examples of these quantities both in pure water and in aqueous solutions are presented in Sections 2.11 through 2.14).

We consider first the Gibbs energy of solvation ΔG_w^* of a water solvaton in a system characterized by the variables T, P, N_w, and m_s, where N_w is the number of water molecules and $m_s = N_s/N_w$, the ratio of the number of solute and solvent molecules. We now expand ΔG_w^* in a Taylor series in the quantity m_s. To first order we have

$$\Delta G_w^*(T, P, N_w, m_s) = \Delta G_w^{*P}(T, P, N_w) + \frac{\partial \Delta G_w^*}{\partial m_s} m_s + \cdots \quad (1.100)$$

where $\Delta G_w^{*P}(T, P, N_w)$ is the solvation Gibbs energy of a water molecule in pure liquid water at T, P.

The linear coefficient in the above expansion is now denoted by

$$\delta G^*(w/s) = \lim_{m_s \to 0} \left(\frac{\partial \Delta G_w^*}{\partial m_s} \right)_{T,P,N_w} \quad (1.101)$$

$\delta G^*(w/s)$ will be referred to as the limiting effect of the solute s on the solvation Gibbs energy of water.

Using the general expression (1.87) and noting that $\mu_w^E = O(m^2)$ (see Section 3.22), we rewrite (1.101) in the form

$$\delta G^*(w/s) = \lim_{m_s \to 0} \frac{\partial}{\partial m_s} \left[kT \ln \left(\frac{\rho_w^w}{\rho_w^l + \rho_s^l} \right) \right] \quad (1.102)$$

where ρ_w^w is the number density of the pure water and ρ_w^l and ρ_s^l are the number densities of water and of the solute in the solution, l, respectively. Note that

$$\left(\frac{\partial \rho_s^l}{\partial m_s}\right)_{T,P,N_w} = \rho_w^l (1 - \rho_s^l \bar{V}_s), \quad \left(\frac{\partial \rho_w^l}{\partial m_s}\right)_{T,P,N_w} = -(\rho_w^l)^2 \bar{V}_s \quad (1.103)$$

where $\bar{V}_s$ is the partial molar volume of s in the solution; hence the limit in (1.102) may be taken, in which case we obtain

$$\delta G^*(\text{w/s}) = \lim_{m_s \to 0} \frac{\partial}{\partial m_s} [-kT \ln(\rho_w^l + \rho_s^l)]$$

$$= \lim_{m_s \to 0} \frac{-kT\rho_w^l}{\rho_w^l + \rho_s^l} (1 - \rho_w^l \bar{V}_s - \rho_s^l \bar{V}_s) = kT\rho_w^w(\bar{V}_s^0 - V_w^w) \quad (1.104)$$

where $\bar{V}_s^0$ is the partial molecular volume of s at infinite dilution and V_w^w is the volume per molecule of pure water.

Thus in order to evaluate $\delta G^*(\text{w/s})$, all we need are the two experimental quantities $\bar{V}_s^0$ and V_w^w [note that $\rho_w^w = (V_w^w)^{-1}$].

The molecular interpretation of the quantity $\delta G^*(\text{w/s})$ is the following. Suppose we look at the difference

$$\Delta G_w^*(T, P, N_w, N_s = 1) - \Delta G_w^*(T, P, N_w) \quad (1.105)$$

i.e., the effect of *one* s molecule on the solvation Gibbs energy of water. Expansion to first order in N_s yields

$$\Delta G_w^*(T, P, N_w, N_s = 1) - \Delta G_w(T, P, N_w) = \left(\frac{\partial \Delta G^*}{\partial N_s}\right)_{T,P,N_w} + \cdots$$

$$= \frac{kT}{N_w} \left(\frac{\bar{V}_s^0}{V_w^w} - 1\right) \quad (1.106)$$

Clearly, in the limit of the macroscopic system this quantity becomes negligibly small. The reason in simple: adding just one s to a macroscopic system of N_w water molecules would have a negligible effect on the solvation thermodynamics of a water solvaton. However, if we multiply equation (1.106) by N_w, we obtain exactly the quantity $\delta G(\text{w/s})$, which is interpreted as N_w times the effect of one s molecule on the sovation Gibbs energy of a water molecule. This is also the reason we have chosen the variable m_s in the expansion (1.100) to obtain a quantity that is finite in the limit of macroscopic systems.

The process for which expression (1.106) is its Gibbs energy change is depicted in Figure 1.4.

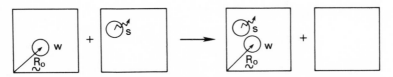

Figure 1.4. The process of transferring one w molecule from a fixed position in pure water to a solution containing one solute s (at constant P, T). The quantity $\delta G^*(w/s)$ is N_w times the Gibbs energy change for this process.

Another molecular interpretation of the quantity $\delta G^*(w/s)$ follows from the Kirkwood–Buff theory of solution.[4,6] Some important results of this theory are presented in Section 3.10. Here we quote only one result for the partial molar (or molecular) volume of s at infinite dilution in w. The result is

$$\bar{V}_s^0 = V_w^w + (G_{ww}^0 - G_{ws}^0) \tag{1.107}$$

where $G_{\alpha\beta}$ are integrals over the corresponding pair correlation function $g_{\alpha\beta}(R)$ for the species α, β, i.e.,

$$G_{\alpha\beta} = \int_0^\infty [g_{\alpha\beta}(R) - 1]4\pi R^2 dR \tag{1.108}$$

and the superscript zero signifies the limit $m_s \to 0$. Substitution of expression (1.107) in equation (1.104) yields

$$\delta G^*(w/s) = kT\rho_w^w(G_{ww}^0 - G_{ws}^0)$$

$$= kT \int_0^\infty [\rho_w^w g_{ww}(R) - \rho_w^w g_{ws}(R)]4\pi R^2 dR \tag{1.109}$$

The quantity $\rho_w^w g_{ww}(R)4\pi R^2 dR$ is the average number of w particles in the spherical shell of width dR at distance R from a w molecule. Hence the integral on the rhs of equation (1.109) is the average excess of water molecules around a water molecule relative to water molecules around an s molecule. In other words, $\delta G^*(w/s)$ is a measure of the relative affinity between the water–water and water–solute pairs.

In a similar manner we may define the limiting effect of s on all the solvation quantities of a water molecule.

For the entropy, we have

$$\delta S^*(w/s) = \lim_{m_s \to 0} \left(\frac{\partial \Delta S_w^*}{\partial m_s}\right)_{T,P,N_w}$$

$$= k\rho_w^w(\bar{V}_s^0 - V_w^w) + kT\rho_w^w \left(\frac{\bar{V}_s^0}{V_w^w}\frac{\partial V_w^w}{\partial T} - \frac{\partial \bar{V}_s^0}{\partial T}\right) \tag{1.110}$$

Similarly, for the enthalpy we have

$$\delta H^*(w/s) = \delta G^*(w/s) + T\delta S^*(w/s)$$

$$= kT^2\rho_w^w \left(\frac{\bar{V}_s^0}{V_w^w} \frac{\partial V_w^w}{\partial T} - \frac{\partial \bar{V}_s^0}{\partial T} \right) \tag{1.111}$$

and for the volume

$$\delta V^*(w/s) = kT\rho_w^w \left(\frac{\partial \bar{V}_s^0}{\partial P} - \frac{\bar{V}_s^0}{V_w^w} \frac{\partial V_w^w}{\partial P} \right) \tag{1.112}$$

Some numerical values for these quantities are presented in Section 2.12. Unfortunately, due to the lack of sufficient experimental data, only a few cases of $\delta S^*(w/s)$ and $\delta H^*(w/s)$ are reported, and no data are reported on $\delta V^*(w/s)$ and higher-order derivatives of $\delta G^*(w/s)$.

Finally, it should be noted that the solvation Gibbs energy of a solute s and the solvation Gibbs energy of the solvent w are related through the Gibbs–Duhem relation (see Section 1.8). This means that the solvation Gibbs energy of s, as well as its concentration dependence, could be used to obtain information on the solvation Gibbs energy of water. However, with the new tool we have defined in this section, we can investigate directly the solvation thermodynamics of water and not indirectly, as might be done through the solvation thermodynamics of the solute molecules.

1.8. RELATIONSHIP BETWEEN THE SOLVATION GIBBS ENERGIES OF A AND B IN A TWO-COMPONENT SYSTEM

Consider a mixture of A and B at some temperature T and pressure P. The Gibbs–Duhem relationship is

$$- SdT + VdP = N_A d\mu_A + N_B d\mu_B \tag{1.113}$$

For variations at fixed T and P, we have

$$x_A \left(\frac{\partial \mu_A}{\partial x_A} \right)_{T,P} + x_B \left(\frac{\partial \mu_B}{\partial x_A} \right)_{T,P} = 0 \tag{1.114}$$

where $x_A = N_A/(N_A + N_B)$ is the mole fraction of A in the mixture.

We now use the general expression for the CP of, say, A in the mixture [equation (1.6)] and in an ideal gas [equation (1.7)] to obtain

$$\frac{\partial \mu_A}{\partial x_A} = \frac{\partial \mu_A^*}{\partial x_A} + kT \frac{\partial \ln \rho_A^l}{\partial x_A} \tag{1.115}$$

Since for ideal gases $\mu_A^{*ig} = -kT \ln q_A$ is independent of x_A, we may include $\partial \mu_A^{*ig}/\partial x_A = 0$ in equation (1.115). Applying the same for the two components, we get from equations (1.114) and (1.115)

$$x_A \frac{\partial \Delta G_A^*}{\partial x_A} + x_B \frac{\partial \Delta G_B^*}{\partial x_A} + x_A kT \frac{\partial \ln \rho_A^l}{\partial x_A} - x_B kT \frac{\partial \ln \rho_B^l}{\partial x_B} = 0 \tag{1.116}$$

This may be simplified by noting that

$$x_A \frac{\partial \ln \rho_A^l}{\partial x_A} = x_A \frac{\partial \ln \rho_A^l}{\partial N_A} \frac{\partial N_A}{\partial x_A} = \rho^l \bar{V}_B \tag{1.117}$$

$$x_B \frac{\partial \ln \rho_B^l}{\partial x_B} = \rho^l \bar{V}_A \tag{1.118}$$

where $\rho^l = \rho_A^l + \rho_B^l$. Hence

$$x_A \frac{\partial \Delta G_A^*}{\partial x_A} + x_B \frac{\partial \Delta G_B^*}{\partial x_A} + kT\rho^l(\bar{V}_B - \bar{V}_A) = 0 \tag{1.119}$$

Thus from a knowledge of ΔG_A^* and its dependence on the composition, we may integrate equation (1.119) to obtain ΔG_B^* at any composition. This integration requires also explicit knowledge of the dependence of $\rho_l(\bar{V}_B - \bar{V}_A)$ on x_A; therefore, the whole procedure may not be practical.

An important limiting result of equation (1.119) is when one component, say A, is very dilute in B. Taking the limit of equation (1.119) for $x_A \to 0$ ($x_B \to 1$), we obtain

$$\lim_{x_A \to 0} \frac{\partial \Delta G_B^*}{\partial x_A} = kT\rho_B^B(\bar{V}_A^0 - V_B^B) \tag{1.120}$$

which is the same as equation (1.104) of the previous section. Indeed

$$\lim_{x_A \to 0} \frac{\partial \Delta G_B^*}{\partial x_A} = \lim_{m_A \to 0} \frac{\partial \Delta G_B^*}{\partial m_A}$$

1.9. HIGHER-ORDER DERIVATIVES OF THE SOLVATION GIBBS ENERGY

In this book we shall present values of the Gibbs energy, entropy, enthalpy, and volume of the solvation process. In principle one can define any other thermodynamic quantity pertaining to the solvation process. Some quantities of potential interest are discussed below. Unfortunately, due to lack of sufficient experimental data these cannot be calculated at present.

(1) *Heat capacity of solvation*. The heat capacity (at constant pressure) of the solvation process is defined by

$$\Delta C_s^* = (\partial \Delta H_s^*/\partial T)_{P,N} \tag{1.121}$$

The evaluation of ΔC_s^* from experimental data requires the second derivative with respect to T of the equilibrium densities, i.e., since

$$\Delta H_s^* = -kT^2 \frac{\partial \ln(\rho_s^{ig}/\rho_s^l)_{eq}}{\partial T}$$

then

$$\Delta C_s^* = -kT^2 \frac{\partial^2 \ln(\rho_s^{ig}/\rho_s^l)_{eq}}{\partial T^2} - 2kT \frac{\partial \ln(\rho_s^{ig}/\rho_s^l)_{eq}}{\partial T} \tag{1.122}$$

(2) *Isothermal compressibility of solvation*. The isothermal compressibility of a system l is defined by

$$\beta_T^l = -\frac{1}{V} \left(\frac{\partial V}{\partial P} \right)_{T,N} \tag{1.123}$$

Thus formally one could have defined the isothermal compressibility of the solvation process as the change in β_T^l corresponding to the transfer of s from a fixed position in an ideal gas to a fixed position in l, i.e.,

$$\Delta \beta_{T,s}^* = \beta_{T,s}^{*l} - \beta_{T,s}^{*ig} \tag{1.124}$$

Since the addition of s to a fixed position in an ideal-gas phase does not change its compressibility, $\Delta \beta_{T,s}^*$ in relation (1.24) is essentially the change of isothermal compressibility of the liquid l upon the addition of s to a fixed position. This quantity seems quite cumbersome to evaluate. Instead, it is simpler to define the pressure dependence of ΔV_s^*, i.e., since

ΔV_s^* is already an intensive quantity, we define the quantity

$$\Delta \eta_s^* = \frac{\partial \Delta V_s^*}{\partial P} = kT \frac{\partial^2 \ln(\rho_s^{ig}/\rho_s^l)_{eq}}{\partial P^2} \tag{1.125}$$

The evaluation of this quantity requires knowledge of the second derivative with respect to pressure of the ratio of the densities at equilibrium.

(3) *Thermal expansion coefficient of solvation.* The isobaric thermal expansibility of the phase l is defined by

$$\alpha_P^l = \frac{1}{V} \left(\frac{\partial V}{\partial T} \right)_{P,N} \tag{1.126}$$

As before, one could have formally written the thermal expansibility for a system with and without an s molecule at a fixed position and then formed the difference between an ideal gas and the liquid phase, i.e.,

$$\alpha_{P,s}^{*l} = \alpha_P^l(\mathbf{R}_s) - \alpha_P^l \tag{1.127}$$

$$\alpha_{P,s}^{*ig} = \alpha_P^{ig}(\mathbf{R}_s) - \alpha_P^{ig} \tag{1.128}$$

and

$$\Delta \alpha_{P,s}^* = \alpha_{P,s}^{*l} - \alpha_{P,s}^{*ig} \tag{1.129}$$

Again, we believe it is simpler to evaluate the temperature coefficient of the solvation volume rather than the quantity $\Delta \alpha_{P,s}^*$, i.e.,

$$\Delta \varepsilon_s^* = \frac{\partial \Delta V_s^*}{\partial T} = \frac{\partial^2 [kT \ln(\rho_s^{ig}/\rho_s^l)_{eq}]}{\partial T \partial P} \tag{1.130}$$

This quantity requires knowledge of the second derivative of the ratio of densities with respect to P and T.

Chapter 2

Some Specific Systems

2.1. INTRODUCTION

This chapter is devoted to the presentation of numerical values for the solvation thermodynamics of some specific systems. The emphasis is on the wide range of systems for which the concept of solvation may be applied, ranging from hard spheres to aqueous protein solutions. We note again that this range is infinitely wider than the range of systems that can be studied by the conventional concept of solvation. Thus the systems treated in Sections (2.4), (2.11), and (2.15) are totally inaccessible, whereas the systems treated in Sections (2.5), (2.12), (2.13), (2.14), and (2.17) are almost entirely inaccessible from the traditional approach. In a few cases a qualitative interpretation of various results and trends is attempted.

We start in the next section with the simplest, nontrivial case of solvation in a condensed phase. Then we proceed to more and more complex systems. Of course, at some point the "order of complexity" becomes meaningless. Solvation of a hydrocarbon molecule in pure hydrocarbon liquid is not less, or more, complex than the solvation of the same molecule in liquid water. The chapter ends, however, with a fairly complex system, aqueous protein solutions. Throughout the chapter the selection of the particular systems has been quite arbitrary. No attempt was made to be exhaustive. Neither can we claim that a particular set of data, chosen in one of the sections, is the best of that kind.

2.2. SOLVATION OF A HARD-SPHERE PARTICLE IN A LIQUID CONSISTING OF HARD-SPHERE PARTICLES

We begin this chapter with the simplest, nontrivial solvation pheno-menon, the solvation of a hard-sphere (HS) particle in a fluid of HS particles. Although HSs are not real particles, they serve as a useful and important theoretical idealization of real particles.

A HS particle is characterized by two molecular parameters, its diameter σ and mass m. The diameter is defined in terms of the pair potential operating between two HSs, namely

$$U(R) = \begin{cases} 0 & \text{for } R > \sigma \\ \infty & \text{for } R \leq \sigma \end{cases} \tag{2.1}$$

The interaction energy is zero whenever the interparticle distance is larger than the diameter σ and becomes infinity when $R \leq \sigma$, i.e., the two particles cannot penetrate into each other beyond the distance σ.

Similarly, two HSs of different diameters, σ_A and σ_B, are characterized by a pair potential of the form

$$U_{AB}(R) = \begin{cases} 0 & \text{for } R > \sigma_{AB} \\ \infty & \text{for } R \leq \sigma_{AB} \end{cases} \tag{2.2}$$

where $\sigma_{AB} = (\sigma_A + \sigma_B)/2$.

Consider now a mixture of HSs of two kinds, A and B, with diameters σ_A and σ_B, respectively. The CP of the A particles is

$$\mu_A = \mu_A^* + kT \ln \rho_A \Lambda_A^3 = W(A \mid A + B) + kT \ln \rho_A \Lambda_A^3 \tag{2.3}$$

Clearly, since the HS particle is devoid of any internal degrees of freedom, the PCP μ_A^* is simply the coupling work $W(A \mid A + B)$ of one A to the system.

Consider next two extreme cases. First, suppose that A is very dilute in B, say one A in a system of pure Bs. In this case it follows from the very definition of a HS that the coupling work $W(A \mid B)$ is exactly equal to the Gibbs energy change in creating a spherical cavity of radius $R_{cav} = (\sigma_A + \sigma_B)/2$ at some fixed position in the system.

The second extreme case is of pure A. In this case the coupling work $W(A \mid A)$ is equal to the Gibbs energy change in creating a cavity of radius $R_{cav} = \sigma_A$ at some fixed position in the system. These two cases are depicted in Figure 2.1.

Two comments regarding the identity between cavity formation and coupling work are now in order. In the first place the concept of cavity in its strict sense is somewhat lost if we have a finite concentration of, say, A in B, or if either A or B are not spheres. If a particle is placed at a fixed position in a mixture of A and B, it creates a cavity of diameter σ_A with respect to an approaching A particle and a cavity of diameter $(\sigma_A + \sigma_B)/2$

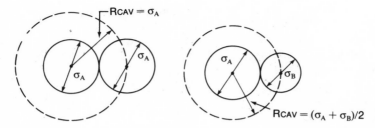

Figure 2.1. Two extreme cases of solvation of a HS A of diameter σ_A. Left: A is solvated by As only; the cavity created by the solvaton has radius $R_{cav} = \sigma_A$. Right: A is solvated by Bs only; the cavity created by the solvaton has radius $R_{cav} = (\sigma_A + \sigma_B)/2$.

with respect to an approaching B particle. Thus, it is somewhat ambiguous to claim that placing an A particle at a fixed position is equivalent to the creation of a "cavity" in the liquid. Such a cavity has no well-defined radius or geometry.

Similarly, if the solvaton A is a HS but the Bs are, say, hard rods, then again it is meaningless to speak of a cavity formed by A in the liquid consisting of Bs. The size of the cavity, or rather the excluded volume, depends on the relative orientation of B toward A. These examples are depicted in Figure 2.2.

The second comment is more important. The equivalency between coupling work and cavity formation (whenever this concept is applicable) is valid *only* for the coupling work, i.e., the PCP. In our approach this is also the same as the solvation Gibbs energy of a HS. This equivalency does not hold for the CP. The CP of a HS fluid is (at least theoretically) a well-defined quantity. The HS particle is presumed to have a mass m, hence momentum, and density ρ. On the other hand, a cavity does not

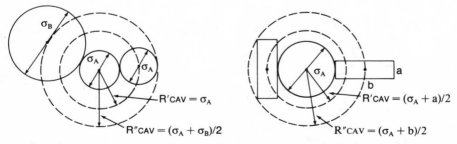

Figure 2.2. A HS of diameter σ_A is solvated by a mixture of As and Bs, with diameters σ_A and σ_B, respectively (left), and by hard rods of length b and cross-section diameter a (right). The dashed circles mark the excluded volumes, or the "cavities" for each approaching particle.

have either momentum or density. Therefore, the liberation Gibbs energy is undefinable for a cavity. In other words, placing a HS at some fixed position is equivalent to a stipulation on the possible distances that other particles may approach the solvaton. This is what we call a cavity. Once we remove the constraint on the fixed position, the identity between the two concepts is lost. The HS gains its liberation Gibbs energy $kT \ln \rho \Lambda^3$, while the cavity cannot be assigned such a liberation Gibbs energy.

As we have noted above, since systems of HS particles do not exist, we cannot rely on experiments to furnish data on their solvation thermodynamics. One possible way of obtaining such information is by computer-simulation techniques. Another theoretical method is the so-called scaled-particle theory (SPT), which is essentially an approximate theory of computing the Gibbs energy of cavity formation in HS systems. Some details on the SPT and references to original articles are given in Section 3.4. Here we present only one typical result of this theory.

Consider a system of one A particle in a "solvent" of B particles. The corresponding diameters are σ_A and σ_B, respectively. The SPT provides a simple recipe for computing the solvation Gibbs energy of the solvaton A, by means of a polynomial of the form

$$W(A \mid B) = K_0 + K_1 R_{cav} + K_2 R_{cav}^2 + K_3 R_{cav}^3 \qquad (2.4)$$

where $R_{cav} = (\sigma_A + \sigma_B)/2$, and the coefficients K_i are given explicitly in Section 3.4.

Figure 2.3 shows the dependence of the reduced solvation Gibbs energy $W(A \mid B)/kT$ on the reduced radius of the cavity $R' = R_{cav}/\sigma_B = (\sigma_A + \sigma_B)/2\sigma_B$.

Note that for very small cavities $R_{cav} \leq \sigma_B/2$ (this is equivalent to a HS of diameter $\sigma_A \leq 0$, or equivalently $R' \leq \frac{1}{2}$), we have an exact expression for the coupling work which reads (or details, see Section 3.4)

$$W(A \mid B) = -kT \ln(1 - \rho_B 4\pi R_{cav}^3/3) \qquad (2.5)$$

where ρ_B is the solvent density and $R_{cav} \leq \sigma_B/2$. In Figure 2.3 the full points correspond to equation (2.5), while the open circles were computed by the polynomial (2.4).

It should be noted that in this case the solvation Gibbs energy is always positive and that it increases monotonically as a function of the radius of the cavity. In Section 3.4 we shall prove that such a monotonic behavior is not only a result of the SPT, but follows from a more general argument based on the definition of the solvation Gibbs energy of a HS solvaton in a HS fluid.

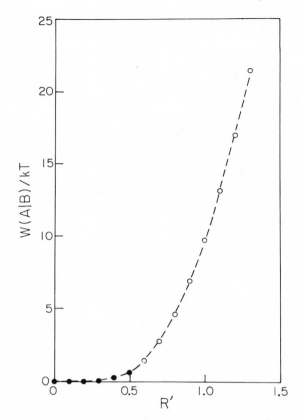

Figure 2.3. Solvation Gibbs energy of a HS "solute" in a HS "solvent" as a function of the reduced radius of the cavity created by the solute in the solvent, $R' = (\sigma_A + \sigma_B)/2\sigma_B$. Values of $W(A \mid B)/kT$ were computed from equation (2.5) (for $R' \leq \frac{1}{2}$) and from equation (2.4) (for $R' \geq \frac{1}{2}$).

2.3. SOLVATION OF A HARD-SPHERE PARTICLE IN A REAL FLUID

A simple generalization of the concept of solvation of a HS particle in a HS fluid can be applied to the case of a real fluid. Here, one needs to assign an effective HS diameter to the real molecules. Then the solvation of a HS in such a fluid, or e uivalently creating a suitable cavity in the fluid, becomes a well-defined process.

For simple spherical molecules, such as the noble gases, a natural choice of the effective diameter might be the van der Waals or Lennard-Jones diameter of the molecules. For more complex molecules there is no universal way of defining an effective HS diameter to be assigned to the solvent molecules.

In spite of this ambiguous concept of the effective HS diameter, a great amount of work has been carried out using the methods of the SPT to calculate the solvation thermodynamics of a HS in various real fluids, ranging from simple noble gases to water and even to ionic melts.

Basically, the SPT is used in these cases to compute the cavity work as a first step in the solvation process. The second step consists of turning on the soft part of the interaction between the solvaton and its entire environment. The formal split of the solvation process into these two parts is discussed in Section 3.5. Both parts of the solvation Gibbs energy are computed by methods of statistical mechanics using molecular parameters as the input data. We shall not review these approaches in the present chapter, however.

2.4. SOLVATION OF AN INERT GAS MOLECULE IN ITS OWN PURE LIQUID

The simplest real molecules for which the solvation quantities can be measured are the inert gases. Two kinds of experimental data that we can use to evaluate solvation quantities are available for these systems. First are the vapor–liquid densities of the two phases along the coexistence equilibrium line. Second are PVT data on inert gas liquids. Both are used in the following. Let ρ_s^g and ρ_s^l be the densities of the inert gas s in the gas and liquid phases, respectively. Using our general expression (1.22) of Chapter 1, we have

$$\Delta\Delta G_s^* = \mu_s^{*l} - \mu_s^{*g} = \Delta G_s^{*l} - \Delta G_s^{*g} = W(s \mid l) - W(s \mid g)$$
$$= kT \ln(\rho_s^g/\rho_s^l)_{eq} \tag{2.6}$$

Thus knowledge of the density ratio is sufficient for obtaining the difference in the solvation Gibbs energies of s in the two phases (or equivalently, the Gibbs energy change for the process of transferring s from a fixed position in g to a fixed position in l, at temperature T and corresponding equilibrium pressure P). Since the inert gas molecules are presumed to possess essentially only translational degrees of freedom, $\Delta\Delta G_s^*$ is also equal to the difference in the coupling work of s in the two phases.

In Figure 2.4 we have plotted $\Delta\Delta G_s^*$ as a function of $T - T_{tr}$, where T_{tr} is the triple point temperature for each of the gases neon, argon, krypton, and xenon. The computations are based on data from Cook,[7] Fastovskii et al.,[8] and Rowlinson and Swinton.[5]

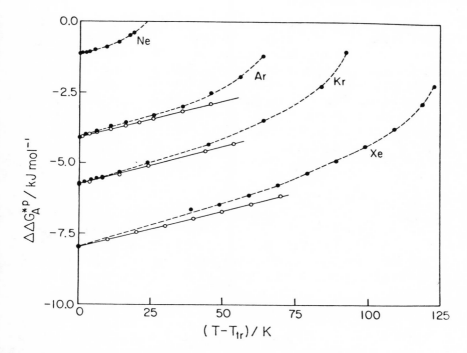

Figure 2.4. Values of $\Delta\Delta G_A^*/\text{kJ mol}^{-1}$ as a function of $(T - T_{tr})/K$ for the inert gases in their own liquids. The triple points for neon, argon, krypton, and xenon were taken as 24.55, 83.78, 115.95, and 161.3 K, respectively. The full points correspond to computations based on equation (2.6), and the open circles to computations based on equation (2.9).

We note that as $T \to T_{tr}$, the density of s in the gaseous phase becomes very low. In this region one can assume that the gaseous phase is an ideal gas, i.e., that $\Delta G_s^{*g} = W(s \mid g) \approx 0$ and hence $\Delta\Delta G_s^*$ is simply the solvation free energy of s in the liquid phase (near $T = T_{tr}$ and at the corresponding low pressure). Thus we always have

$$\Delta G_s^{*l} = \lim_{T \to T_{tr}} \Delta\Delta G_s^* \qquad (2.7)$$

Another limiting case is the critical point. As we approach the critical temperature T_{cr}, the densities of s in the two phases approach each other; hence

$$\lim_{T \to T_{cr}} \Delta\Delta G_s^* = 0 \qquad (2.8)$$

The tendency is clearly discernible for all the gases in Figure 2.4.

Another source of data that we have used, only for the low-pressure region, is the *PVT* data published by Theeuwes and Bearman.[9] Here, both the density of the liquid phase and the total pressure of the system are presented as an analytical function of the temperature.

In order to convert these data into solvation thermodynamics, we require the densities of the gas in the gaseous phase. We therefore assumed that for pressures up to 1 atm the gaseous phase is ideal. Hence we compute the solvation Gibbs energy directly from the available data, i.e.,

$$\Delta G_s^{*l} = kT \ln(P/kT\rho_s^l)_{eq} \tag{2.9}$$

In Figure 2.4 we have added the open circles, which were computed by equation (2.9). The corresponding data for argon were taken from Rowlinson and Swinton.[5] As expected, the two sets of data converge in the limit $T \to T_{tr}$ where the gaseous phase becomes ideal. Clearly, as $T - T_{tr}$ increases, $\Delta\Delta G_s^*$ is no longer equal to ΔG_s^{*l} and equation (2.9) is no longer valid; hence the discrepancies between the two sets of curves diverge.

In Table 2.1 we summarize the values of $\Delta\Delta G_s^*$ for the five inert gases as a function of T above T_{tr}.

In order to compute the entropy and enthalpy of solvation, we need the temperature derivative of $\Delta\Delta G_s^*$ at constant pressure. The data available, though, are along the equilibrium line. The two derivatives are related to each other by

$$\left(\frac{d\Delta\Delta G_s^*}{dT}\right)_{eq} = \left(\frac{\partial\Delta\Delta G_s^*}{\partial T}\right)_P + \left(\frac{\partial\Delta\Delta G_s^*}{\partial P}\right)_T \left(\frac{dP}{dT}\right)_{eq} \tag{2.10}$$

Both of the straight derivatives (taken along the equilibrium line) may be computed from the available experimental data, provided we can assume ideal-gas behavior in the gaseous phase. Hence from now on we assume that $\Delta\Delta G_s^*$ is essentially ΔG_s^*, so equation (2.10) reduces to

$$\left(\frac{d\Delta G_s^*}{dT}\right)_{eq} = -\Delta S_s^* + \Delta V_s^* \left(\frac{dP}{dT}\right)_{eq} \tag{2.11}$$

where ΔS_s^* and ΔV_s^* are the entropy and volume of solvation of s in the pure liquid s, taken in the limit of low pressure (or equivalently, using the experimental data near $T \sim T_{tr}$).

Clearly, equation (2.11) cannot be solved for both ΔS_s^* and ΔV_s^*. Fortunately, however, ΔV_s^* may be computed independently from the

Table 2.1
Values of $\Delta\Delta G_s^*$/kJ mol^{-1} as a Function of T/K for the Inert Gases in Their Own Liquid along the Saturation Curve[a]

Helium-3		Neon		Argon		Krypton		Xenon	
T	$\Delta\Delta G_s^*$	T	$\Delta\Delta G_s^*$	T	$\Delta\Delta G_s^*$	T	$\Delta\Delta G_s^*$	T	$\Delta\Delta G_s^*$
1.0	−0.04115	24.55	−1.1435	83.78	−4.0849	116	−5.7326	161.4	−7.9906
1.2	−0.04410	25	−1.1419	84	−4.0774	118	−5.6698	200	−6.5985
1.4	−0.04611	26	−1.1226	86	−4.0128	120	−5.6120	210	−6.4464
1.6	−0.04736	27	−1.0999	88	−3.9569	122	−5.5682	220	−6.1414
1.8	−0.04778	28	−1.0762	90	−3.8507	124	−5.5166	230	−5.7804
2.0	−0.04744	30	−1.0187	95	−3.7133	130	−5.3201	240	−5.3553
2.2	−0.04631	34	−0.8918	100	−3.5640	140	−5.0074	250	−4.9158
2.4	−0.04437	38	−0.7288	110	−3.3206	160	−4.3799	260	−4.4143
2.6	−0.04149	42	−0.4889	120	−2.9889	180	−3.5227	270	−3.7560
2.8	−0.03747	43	−0.4084	130	−2.5041	200	−2.3185	280	−2.8456
3.0	−0.03191			140	−1.9305	208	−1.0582	285	−2.2025
3.1	−0.02828			148	−1.1854				
3.2	−0.02375								
3.3	−0.01766								
3.33	0.0								

[a] Data from Ben-Naim.[11]

molar volume of s (V_s^l) and the compressibility β_T^l of the pure liquid s using the relation (Section 1.4)

$$\Delta V_s^* = V_s^l - kT\beta_T^l \qquad (2.12)$$

The full computational procedure could be carried out only for argon, and we also added methane for comparison. The data were taken from Rowlinson and Swinton.[5]

First, we compute ΔV_s^* using equation (2.12), then we use ΔV_s^* in combination with $(dP/dT)_{eq}$ in equation (2.11) to estimate ΔS_s^*. Finally, we compute $\Delta H_s^* = T\Delta S_s^* + \Delta G_s^*$.

Table 2.2 shows some values of ΔV_s^* and ΔS_s^*, as well as some intermediary values used in the process of computations for argon and methane.

In order to compute values of ΔS_s^*, ΔH_s^*, and ΔV_s^* for the other gases, we first limit ourselves to the region close to the triple point, and due to the lack of a complete set of experimental data we undertake the following approximate procedure. Table 2.2 shows that values of $kT\beta_T^l$ are quite small relative to V_s^l or ΔV_s^*. Also, the term $\Delta V_s^*(dP/dT)$ is quite small compared with $(d\Delta G^*/dT)$. Therefore, for those gases for which we do not have the experimental data on β_T^l and on $(dP/dT)_{eq}$, we assume that we can neglect the second term on the rhs of equation (2.11). This is the same as making $-\Delta S_s^*$ equal to $(d\Delta G_s^*/dT)_{eq}$.

In Table 2.3 we present the thermodynamic quantities of solvation of some inert gases, including methane, near the triple point for each gas. The figures in brackets are approximate values obtained by either neglecting $kT\beta_T^l$ in equation (2.12) or neglecting $\Delta V_s^*(dP/dT)$ in equation (2.11).

A glance at Table 2.3 shows that all values of ΔG_s^* are negative and their absolute magnitude increases with the molecular mass of the molcules. This is in contrast to what we have seen for the solvation Gibbs energy in the HS systems. Clearly, the negative sign is due to the attractive part of the intermolecular potential. Also, it is evident from this table that $T\Delta S_s^*$ values are relatively small compared with ΔH_s^*. This means that the energetic factor, rather than the entropic factor, is dominating ΔG_s^*, i.e., the sign of ΔG_s^* is determined by ΔH_s^*, which is almost identical with ΔE_s^*. (The $P\Delta V_s^*$ term is usually negligible for most of the systems discussed in this book. Some numerical examples are shown in Section 3.21.)

In turn, ΔE_s^* is equal to half the average binding energy of s to the system (see Chapter 5 in Ben-Naim[4]), i.e.,

$$\Delta H_s^* \approx \Delta E_s^* = \tfrac{1}{2}\langle B_s \rangle \qquad (2.13)$$

Table 2.2
Various Factors Involved in the Computation of ΔS_s^* for Argon and Methane[a]

T/K	ρ_s^g (mol/cm³)	ρ_s^l (mol/cm³)	ΔG_s^* (kJ mol⁻¹)	$\Delta\Delta G_s^*/\Delta T$ (J mol⁻¹ K⁻¹)	$kT\beta_T^l$ (cm³/mol)	ΔV_s^* (cm³/mol)	$\Delta V_s^*(dP/dT)$ (J mol⁻¹ K⁻¹)	ΔS_s^* (J mol⁻¹ K⁻¹)
Argon								
83.80	9.889×10^{-5}	0.03540	−4.0974	26.5	1.382	26.858	0.2156	−26.284
85	1.1179×10^{-4}	0.03522	−4.0656	26.228	1.437	26.953	0.2409	−25.987
87.28	1.396×10^{-4}	0.03485	−4.0058	25.551	1.547	27.143	0.2937	−25.257
90	1.7882×10^{-4}	0.03443	−3.9363	25.34	1.698	27.342	0.3664	−24.974
95	2.7057×10^{-4}	0.03364	−3.8096	25.08	2.057	27.663	0.5256	−24.554
100	3.9056×10^{-4}	0.03282	−3.6842	25.14	2.554	27.916	0.7230	−24.417
105	5.4242×10^{-4}	0.03196	−3.5585	25.44	3.222	28.068	0.9571	−24.483
110	7.2881×10^{-4}	0.03104	−3.4313	26.52	4.097	28.113	1.2286	−25.291
120	1.2159×10^{-3}	0.02904	−3.1661	29.42	7.132	27.298	1.8345	−27.586
Methane								
90.68	1.5546×10^{-5}	0.02813	−5.6551	27.6394	1.08	34.47	0.05205	−27.587
100	4.1474×10^{-5}	0.02736	−5.3975	26.3900	1.413	35.137	0.12615	−26.264
110	9.6653×10^{-5}	0.02648	−5.1336	25.8120	1.911	35.849	0.26924	−25.543
111.631	1.0917×10^{-4}	0.02634	−5.0915	25.4958	2.012	35.958	0.30026	−25.196
120	1.9235×10^{-4}	0.02555	−4.8781	25.3200	2.633	36.497	0.49563	−24.824
130	3.4049×10^{-4}	0.02457	−4.6249	25.7100	3.737	36.963	0.81692	−24.893
140	5.515×10^{-4}	0.02350	−4.3678	26.7700	5.476	37.064	1.23060	−25.539
150	8.3396×10^{-4}	0.02233	−4.1001	29.0200	8.478	36.302	1.70620	−27.314

[a] Data from Ben-Naim.[11]

where $\langle B_s \rangle$ is an ensemble average of the binding energy B_s of s. For more details, see Section 3.8.

From equation (2.13) we may also compute, in a very crude way, the coordination number of s molecules around s. To do that, we assume that $\langle B_s \rangle$ may be approximated by a product of the average coordination number and the value of the interaction energy between two s molecules at the distance of the minimum of the intermolecular potential. If we take for the latter the Lennard-Jones energy parameters, ε_{LJ} (as determined from second virial coefficients[10]), we write the approximation

$$\langle B_s \rangle = - \langle CN \rangle \varepsilon_{LJ} \tag{2.14}$$

Thus from relations (2.13) and (2.14) we have computed approximate values of the average coordination numbers $\langle CN \rangle$ for the various gases in the liquid state, near their triple point. These are shown in Table 2.4. The

Table 2.3

Values of $\Delta G_s^*/kJ\ mol^{-1}$, $T\Delta S_s^*/kJ\ mol^{-1}$, $\Delta H_s^*/kJ\ mol^{-1}$, and $\Delta V_s^*/cm^3\ mol^{-1}$ for Neon, Argon, Krypton, Xenon, and Methane near Their Corresponding Triple Points

	T_{tr}/K	ΔG_s^*	$T\Delta S_s^*$	ΔH_s^*	ΔV_s^*
Neon	24.55	−1.144	[−0.44]	[−1.58]	[16.18]
Argon	83.80	−4.097	−2.202	−6.299	26.86
Krypton	115.95	−5.733	[−2.66]	[−8.53]	[34.22]
Xenon	161.3	−7.991	[−3.94]	[−11.89]	[44.29]
Methane	90.68	−5.6551	−2.502	−8.157	34.47

[a] Approximate values, computed as described in Section 2.4, are included in brackets. Data from Ben-Naim.[11]

Table 2.4

Lennard-Jones Parameters and Average Coordination Numbers $\langle CN \rangle$ of the Inert Gases in Their Liquids at the Triple Points

	$\varepsilon_{LJ}/kJ\ mol^{-1}$	$\langle CN \rangle$
Ne	0.290	10.9
Ar	0.996	12.6
Kr	1.422	12.0
Xe	1.837	12.9
CH_4	1.232	13.2

values obtained are all between 11 and 13, which are quite consistent with what we should expect from a nearly close packing of spheres in the liquid state. For more details on the thermodynamics of solvation in these systems, the reader is referred to the original article.[11]

Finally, we comment on the wel-known Trouton's law in the light of the solvation entropies of inert gases. The general expression for the entropy of condensation may be derived from the partial molecular entropies of s in the two phases, i.e.,

$$-\bar{S}_s^l = \partial\mu_s^l/\partial T = -S_s^{*l} + k\ln\rho_s^l\Lambda_s^3 + kT\partial(\ln\rho_s^l\Lambda_s^3)/\partial T \quad (2.15)$$

$$-\bar{S}_s^g = \partial\mu_s^g/\partial T = -S_s^{*g} + k\ln\rho_s^g\Lambda_s^3 + kT\partial(\ln\rho_s^g\Lambda_s^3)/\partial T \quad (2.16)$$

For the entropy of transfer of one molecule from the gas into the liquid, we have the general expression

$$\bar{S}_s^l - \bar{S}_s^g = S_s^{*l} - S_s^{*g} + k\ln(\rho_s^g/\rho_s^l) + kT(\alpha_P^l - \alpha_P^g) \quad (2.17)$$

where α_P^l is the isobaric thermal expansion of the phase l. Assuming that the gas is ideal, so that $\alpha_P^g = T^{-1}$ and s is at equilibrium between the two phases, equation (2.17) reduces to

$$\Delta S_{\text{cond}} = \bar{S}_s^l - \bar{S}_s^g = \Delta S_s^{*l} + k\ln(\rho_s^g/\rho_s^l)_{\text{eq}} + kT\alpha_P^l - k \quad (2.18)$$

We now evaluate ΔS_{cond} for argon and methane at their normal boiling points $T_b = 87.3$ K and $T_b = 111.63$ K, respectively, and list the values of the four terms in the same order as on the rhs of equation (2.18). Thus we have

$$\Delta S_{\text{cond}}(\text{argon at } T_b)/\text{J mol}^{-1}\text{K}^{-1}$$

$$= -25.26 - 45.6 + 3.27 - 8.314 = -75.93 \quad (2.19)$$

$$\Delta S_{\text{cond}}(\text{methane at } T_b)/\text{J mol}^{-1}\text{K}^{-1}$$

$$= -25.26 - 45.6 + 3.18 - 8.314 = -75.99 \quad (2.20)$$

We see that the dominating quantities in equation (2.18) are the terms $k\ln(\rho_s^g/\rho_s^l)_{\text{eq}}$ and ΔS_s^{*l}, which are probably responsible for the almost constancy of the entropy of condensation for many liquids.[5] It is of interest to note that in 1915, Hildebrand[12] proposed a modification of Trouton's law. Instead of listing values of the entropies of condensation of liquids at their normal boiling point, he suggested looking at the condensation entropies of the gases at temperatures at which the concentration of their vapors are equal. In view of equation (2.18), this modification amounts to the choice of a constant value of ρ_s^g. Hence we expect that fixing ρ_s^g would make the variation in $k \ln(\rho_s^g/\rho_s^l)$ more moderate than when determined at the normal boiling point. This has indeed been observed by Hildebrand.[12]

2.5. SOLVATION OF INERT GAS MOLECULES IN INERT GAS–LIQUID MIXTURES

In the previous section we computed some thermodynamic quantities of solvation of inert gases in their own liquids. We now follow the variation in the solvation quantities when we add a second component to the system. In this section the second component is also an inert gas liquid.

The two-component inert gas mixtures, for which there seem to be quite abundant experimental data, are the argon–krypton system[13,14] and, with somewhat less data, the krypton–xenon mixtures.[15] For these systems the excess Gibbs energies and excess volumes (per mole of mixture) are given as a function of the mole fraction, of one of the components, in an analytical form as follows:

$$g^E = x_A x_B RT[A + B(x_A - x_B) + C(x_A - x_B)^2] \tag{2.21}$$

$$v^E = x_A x_B[A^1 + B^1(x_A - x_B) + C^1(x_A - x_B)^2] \tag{2.22}$$

For the argon–krypton system the coefficients in equation (2.21) are available at two temperatures (103.94 K and 115.77 K). This enables us to compute approximate values for the entropy and enthalpy of solvation in these mixtures (for more details, see Ben-Naim[11]). On the other hand, for the krypton–xenon system, the data available are at one temperature only (161.36 K).

From equations (2.21) and (2.22) we can compute the solvation Gibbs energies of both components relative to the solvation Gibbs energies in the pure state, according to the procedure described in Section 1.5. For instance, for the A component in a mixture of A and B, we have

$$\Delta\Delta G_A^* = \Delta G_A^*(\text{in mixture}) - \Delta G_A^{*p}$$

$$= kT \ln(\rho_A^p x_A/\rho_A) + g^E + x_B(\partial g^E/\partial x_A)$$

$$= kT \ln\left(\frac{v^E + x_A V_A^p + x_B V_B^p}{V_A^p}\right) + g^E + x_B(\partial g^E/\partial x_A) \tag{2.23}$$

and a similar equation holds for the second component B. Figure 2.5 shows the variation of $\Delta\Delta G_A^*$ for both argon and krypton in the argon–

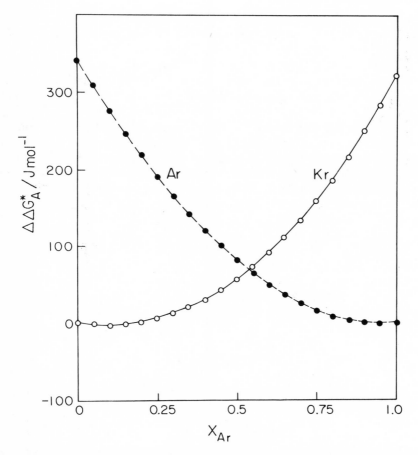

Figure 2.5. Values of $\Delta\Delta G_A^*/\mathrm{J\,mol}^{-1}$ for argon and krypton as a function of the mole fraction of argon at $T = 115.77$ K.

krypton mixtures at the one temperature $T = 115.77$ K. Similarly, the values of $\Delta\Delta G_A^*$ for krypton and xenon in the krypton–xenon mixtures are shown in Figure 2.6 at $T = 161.38$ K.

Also, the relative entropy and enthalpy of solvation of argon and krypton were estimated by taking the difference between two temperatures, i.e.,

$$\Delta\Delta S_A^* = -\partial(\Delta\Delta G_A^*/\partial T)$$

$$\approx -\frac{\Delta\Delta G_A^*(115.77\text{ K}) - \Delta\Delta G_A^*(103.94\text{ K})}{115.77 - 103.94} \tag{2.24}$$

$$\Delta\Delta H_A^* = \Delta\Delta G_A^*(115.77\ K) + 115.77\ \Delta\Delta S_A^*(115.77\ K) \qquad (2.25)$$

Values of $\Delta\Delta G_A^*$, $T\Delta\Delta S_A^*$, and $\Delta\Delta H_A^*$ as a function of the mole fraction of argon are shown for argon and krypton in Figures 2.7 and 2.8, respectively. Some selected values of these quantities are also given in Table 2.5.

As expected, the curves for $\Delta\Delta G_A^*$ approach zero when the composition of the system approaches that of pure A. Less expected is the finding that all the curves of $\Delta\Delta G_A^*$ in Figures 2.5 and 2.6 are almost monotonically decreasing functions of the corresponding mole fractions. (This is true except for a small region at low concentrations, where $\Delta\Delta G_A^*$

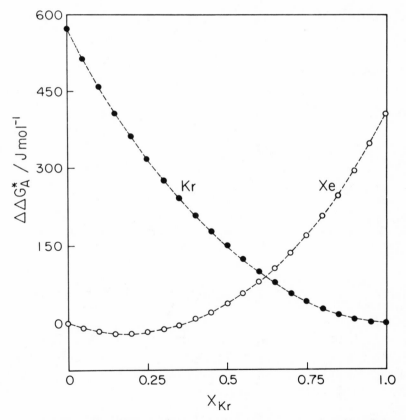

Figure 2.6. Values of $\Delta\Delta G_A^*/J\ mol^{-1}$ for krypton and xenon as a function of the mole fraction of krypton at $T = 161.38\ K$.

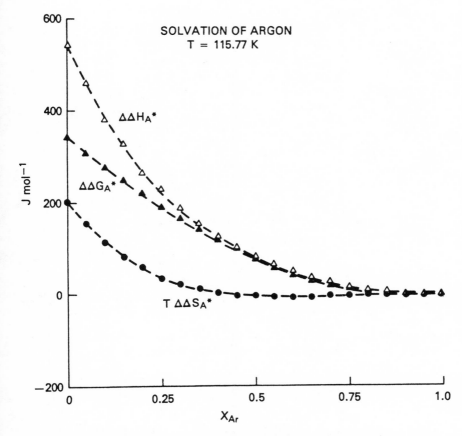

Figure 2.7. Values of $\Delta\Delta G_A^*$, $\Delta\Delta H_A^*$, and $T\Delta\Delta S_A^*$ for argon in mixtures of argon and krypton at $T = 115.77$ K as a function of the mole fraction of argon.

are negative. We do not know if these small negative values are real or an artifact of the curve fitting employed by the authors to the experimental data.)

The fact that all curves reach a positive value when the corresponding component becomes very dilute is indeed quite surprising. Let us examine in some detail the case of the argon–krypton system. First, we note that the main contribution to the Gibbs energy change is due to the enthalpy change rather than the entropy change. This is quite clear from Figures 2.7 and 2.8. Therefore, in the following we shall assume that $\Delta\Delta G_s^*$ is determined essentially by a product of an average interaction energy parameter and an average coordination number.

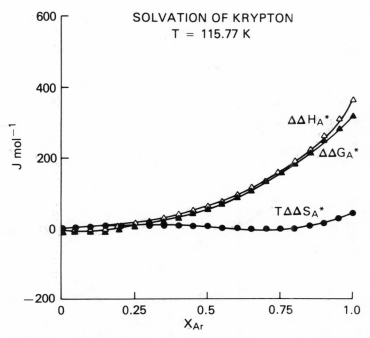

Figure 2.8. Values of $\Delta\Delta G_A^*$, $\Delta\Delta H_A^*$, and $T\Delta\Delta S_A^*$ for krypton in mixtures of argon and krypton at $T = 115.77$ K as a function of the mole fraction of argon.

Table 2.5
Values of $\Delta\Delta G_A^*$, $T\Delta\Delta S_A^*$, and $\Delta\Delta H_A^*$
(all in J mol^{-1}) for Argon in Mixtures of
Argon and Krypton at $T = 115.77$ K^a

x_A	$\Delta\Delta G_A^*$	$T\Delta\Delta S_A^*$	$\Delta\Delta H_A^*$
0	342.6	143.7	486.3
0.1	277.8	68.6	346.2
0.2	219.1	22.3	241.4
0.3	166.6	− 2.5	164.1
0.4	120.7	− 12.3	108.4
0.5	81.5	− 12.5	69.0
0.6	49.5	− 7.9	41.6
0.7	24.9	− 2.0	22.9
0.8	8.2	2.3	10.5
0.9	− 0.2	3.3	3.1
1.0	0.0	0	0

^{a} x_A is the mole fraction of argon. Data from Ben-Naim.[11]

We consider first the case of argon as a solvaton in pure argon. Let ε_{AA} be the interaction parameter and CN_{AA} the average coordination number of argon in pure argon. Changing x_A from 1 to 0 has the effect of replacing all the argon molecules in the surroundings of the solvaton by krypton molecules. This is depicted schematically in Figure 2.9. In this extreme dilution of argon in krypton, the solvation Gibbs energy is determined essentially by a product of the energy parameter ε_{AK} and average coordination number CN_{AK}. Thus for $\Delta\Delta G_A^*$ at $x_A \to 0$ we may write approximately

$$\Delta\Delta G_A^* \approx \Delta\Delta H_A^* \approx -\varepsilon_{AK}CN_{AK} + \varepsilon_{AA}CN_{AA} \qquad (2.26)$$

In this case we expect that $\varepsilon_{AK} > \varepsilon_{AA}$; we also expect that $CN_{AA} > CN_{AK}$ (i.e., packing of Ar around Ar is more efficient than packing of Kr around Ar; see Figure 2.9). This means that the two factors, energy and coordination number, change in opposite directions. The experimental finding that $\Delta\Delta G_A^*$ (of argon at $x_A \to 0$) is positive indicates that the coordination number *is* the factor responsible for the sign of $\Delta\Delta G_A^*$ in relation (2.26) (i.e., ΔG_A^* decreases in absolute magnitude when changing from pure Ar into pure Kr).

The situation seems to be reversed in the case of the solvation of krypton. Here again we write, similar to relation (2.26), the

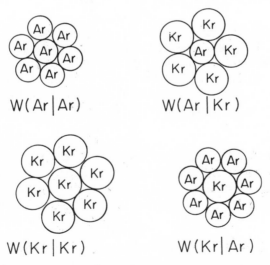

Figure 2.9. Schematic illustration of an argon solvaton, solvated by argon and krypton molecules (above). Similar change of the environment of a krypton solvaton is shown (below). The correspondent coupling work is indicated in each case.

approximation

$$\Delta\Delta G_K^* \approx \Delta\Delta H_K^* \approx -\varepsilon_{KA}CN_{KA} + \varepsilon_{KK}CN_{KK} \qquad (2.27)$$

In this case we expect that $\varepsilon_{KK} > \varepsilon_{KA}$ but $CN_{KA} > CN_{KK}$ (see Figure 2.9). Thus when we change the composition from $x_K = 1$ to $x_K = 0$, the energy parameter decreases while the coordination number increases. Again, the two factors change in opposite directions. The experimental finding that $\Delta\Delta G_K^*$ is positive at $x_K \to 0$ indicates that in this case it is the change in the energy parameter rather than the coordination number which is responsible for the sign of $\Delta\Delta G_K^*$ (at $x_K \to 0$).

From the above analysis, we see that it is not possible, even qualitatively, to predict how the solvation Gibbs energy will change when the solvating liquid is changed from one pure component to another.

In more general cases, ΔG_s^* is determined by both the enthalpy and entropy of solvation. We have allowed ourselves to ignore the changes in entropy, based on the findings of Figures 2.7 and 2.8. This is not always the case. As we shall see in Section 2.9, the solvation of, say, argon in water is determined by the entropy rather than the energy of solvation. In such systems, a simple analysis based on energy parameter and coordination number is clearly inappropriate.

2.6. SOLVATION OF XENON IN n-ALKANES

The solvation thermodynamics of inert gases in various liquid hydrocarbons has been investigated by several researchers. Perhaps the most complete and recent study of these systems has been undertaken by Pollack and Himm[16,17] (see also Ben-Naim and Marcus[18]). In this section we present some results based on data taken from Pollack and Himm on the solubilities of xenon in 16 liquid n-alkanes (from n-pentane to n-eicosane).

The primary quantity obtained from the measurements of the solubilities is the Ostwald absorption coefficient of the solute s (here xenon) in various liquids l. From this quantity we can compute directly the solvation Gibbs energy† of xenon in these liquids through relation (1.23), i.e.,

$$\Delta G_s^* = kT\ln(\rho_s^g/\rho_s^l)_{eq} \qquad (2.28)$$

---

† In their original paper Pollack and Himm[17] calculated the values of $\Delta G(x\text{-process})$. However, recently the solvation Gibbs energies were recalculated by Ben-Naim and Marcus.[18]

In this particular case, since the solute is an inert gas molecule and the sovents are relatively inert liquids as well, we may assume that ΔG_s^* is essentially the coupling work of xenon to the liquid [see equation (1.13), Section 1.2]:

$$\Delta G_s^* = W(s \mid n\text{-alkane}) \tag{2.29}$$

The solvation Gibbs energies of xenon in these liquids are plotted as a function of n, the number of carbon atoms in the n-alkane, in Figure 2.10. Also, from the temperature dependence of ΔG_s^* we have computed both ΔS_s^* and ΔH_s^*. Some selected values are given in Table 2.6 at one temperature.

A glance at Table 2.6 reveals that, in all of these cases, the solvation Gibbs energy is dominated by the enthalpy rather than the entropy of solvation. This behavior is similar to that encountered in the previous two sections. Therefore, in a qualitative attempt to interpret the experimental data, we rely mainly on energetic considerations.

A qualitative interpretation of the general trends observed in Figure 2.10 has been suggested,[18] based on the assumption that the interaction energy parameter between xenon and a methyl group $\varepsilon(Xe, CH_2)$ is somewhat stronger than the interaction energy parameter between xenon and a methylene group $\varepsilon(Xe, CH_2)$. As in previous sections, we also

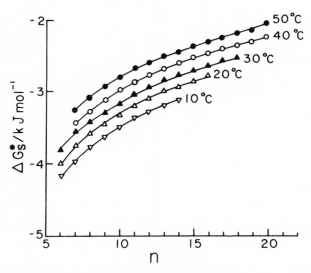

Figure 2.10. Solvation Gibbs energies of xenon in n-alkanes as a function of n, the number of carbon atoms in the n-alkanes.

Table 2.6
Values of ΔG_s^*, $T\Delta S_s^*$, and ΔH_s^*
for Xenon in n-Alkanes at 20 °C^a

Solvent	ΔG_s^*	$T\Delta S_s^*$	ΔH_s^*
n-C_6H_{14}	− 3.953	− 5.27	− 9.23
n-C_7H_{16}	− 3.757	− 5.76	− 9.52
n-C_8H_{18}	− 3.589	− 5.27	− 8.86
n-C_9H_{20}	− 3.464	− 5.15	− 8.61
n-$C_{10}H_{22}$	− 3.330	− 4.90	− 8.24
n-$C_{11}H_{24}$	− 3.200	− 4.78	− 7.98
n-$C_{12}H_{26}$	− 3.112	− 4.90	− 8.02
n-$C_{13}H_{28}$	− 3.012	− 5.14	− 8.16
n-$C_{14}H_{30}$	− 2.945	− 4.90	− 7.85
n-$C_{15}H_{32}$	− 2.866	− 4.66	− 7.53
n-$C_{16}H_{34}$	− 2.786	− 3.92	− 6.71

^{a} All values are in kJ mol^{-1}. Data from Ben-Naim and Marcus.[18]

assume that a xenon atom interacts only with the surrounding molecules in the first coordination sphere. Let $CN(CH_3)$ and $CN(CH_2)$ be the average coordination numbers of methyl and methylene groups around the xenon solvaton in an n-alkane liquid. The enthalpy, or the energy of solvation, of xenon in this liquid may be approximated by

$$\Delta H_s^* \approx \Delta E_s^* \approx - [\varepsilon(Xe, CH_3)CN(CH_3) + \varepsilon(Xe, CH_2)CN(CH_2)]$$

$$= - CN(CH_2)[\varepsilon(Xe, CH_3)CN(CH_3)/CN(CH_2) + \varepsilon(Xe, CH_2)] \quad (2.30)$$

Now we assume the following: first, that the energy parameters fulfill the inequality

$$\varepsilon(Xe, CH_3) > \varepsilon(Xe, CH_2) \quad (2.31)$$

and second, since as n increases the ratio of the number of methylene to methyl groups in the bulk liquid will increase, the same must also be reflected in the ratio $CN(CH_3)/CN(CH_2)$ in the surroundings of the solvaton. Finally, we note that the molar volume of the n-alkane liquids at 20 °C is almost a linear function of n and has the form

$$V_m/cm^3\ mol^{-1} = 33.14 + 16.18n \quad (2.32)$$

Hence the average number densities of methyl and methylene groups will depend on n as

$$\rho_{CH_3} \approx \frac{2}{V_m} = \frac{2}{33.14 + 16.18n} \qquad (2.33)$$

$$\rho_{CH_2} \approx \frac{n-2}{V_m} = \frac{n-2}{33.14 + 16.18n} \qquad (2.34)$$

From the above assumptions it is clear that, as n increases, the xenon–methylene interaction becomes dominant. Therefore, the values of ΔH_s^* are expected to decrease in absolute magnitude, which in turn affects also the general trend of ΔG_s^* as a function of n. This is consistent with the experimental observation as depicted in Figure 2.10. The fact that the absolute values of ΔG_s^* decrease with an increase in temperature is probably due to the expansion of the liquid or, equivalently, to the decrease in the average number of methyl and methylene groups in the surroundings of the xenon atom.

The above interpretation was presented to demonstrate the consistency between the observed values of the solvation Gibbs energy (sign, dependence on n and T) and a qualitative molecular picture of the surroundings of a solvated molecule. This kind of interpretation could not have been done with the conventional Gibbs energy of solvation. In fact, Pollack and Himm found that $\Delta G(x\text{-process})$ are all *positive* and *decrease* with n, but increase with the temperature. Also, they found that for the x-process the entropy rather than the enthalpy is the dominant effect. This is in sharp contrast to our findings regarding the solvation process.

2.7. SOLVATION OF XENON IN n-ALKANOLS

An extension of the work on the solubilities of xenon in n-alkanes to n-alkanols has recently been published by Pollack *et al.*[19] Table 2.7 presents the solvation free energies of xenon in various n-alkanols at different temperatures. In Figure 2.11 we present values of ΔG_s^*, ΔH_s^*, and $T\Delta S_s^*$ for xenon in n-alkanes and n-alkanols at one temperature, 20 °C. The most significant difference is that ΔG_s^* values in n-alkanes are more negative than ΔG_s^* in n-alkanols. This is quite unexpected if we rely only on interaction-energy considerations. Using arguments similar to those of the previous section, we would have expected that addition of a hydroxyl group to an n-alkane would increase the interaction energy between xenon and the solvent. However, we do not really have any information regarding the density of the hydroxyl groups around the

Table 2.7
Values of $\Delta G_s^*/kT$ mol^{-1} of Xenon in n-Alkanols at
Various Temperatures[a]

Solvent	10 (°C)	20 (°C)	30 (°C)	40 (°C)	50 (°C)
CH_3CH	− 2.123	− 1.926	− 1.722	− 1.516	—
C_2H_5OH	− 2.418	− 2.209	− 2.015	− 1.824	− 1.654
n-C_3H_7OH	− 2.599	− 2.375	− 2.181	− 2.004	− 1.830
n-C_4H_9OH	− 2.618	− 2.405	− 2.207	− 2.017	− 1.837
n-$C_5H_{11}OH$	− 2.566	− 2.348	− 2.161	− 1.965	− 1.787
n-$C_6H_{13}OH$	− 2.560	− 2.339	− 2.145	− 1.954	− 1.756
n-$C_7H_{15}OH$	− 2.517	− 2.299	− 2.106	− 1.914	− 1.732
n-$C_8H_{17}OH$	− 2.471	− 2.248	− 2.048	− 1.865	− 1.692
n-$C_9H_{19}OH$	− 2.416	− 2.220	− 2.027	—	—
n-$C_{10}H_{21}OH$	− 2.376	− 2.169	− 1.989	− 1.807	− 1.625
n-$C_{11}H_{23}OH$	—	− 2.076	− 1.882	− 1.700	− 1.522
n-$C_{12}H_{25}OH$	—	—	− 1.897	− 1.719	− 1.546

[a] Based on Data from Pollack *et al.*[19]

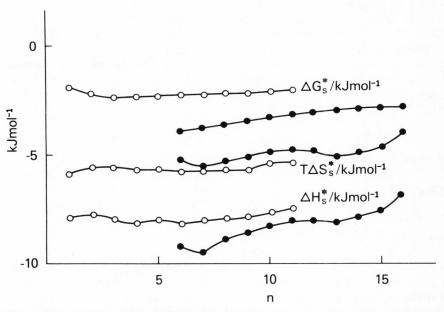

Figure 2.11. Solvation Gibbs energies, entropies, and enthalpies of xenon in n-alkanols (O) and n-alkanes (●) as a function of n, the number of carbon atoms in the solvent molecules at $t = 20$ °C.

xenon atom. Therefore, all we can conclude from the present experimental data is that the average binding energy of xenon to its surrounding is reduced on replacing an environment of *n*-alkane by *n*-alkanol. This reduction is probably due to the decrease in the overall coordination number of the various groups around the xenon atom.

2.8. SOLVATION OF INERT GAS MOLECULES IN SOME ORGANIC LIQUIDS

There exists a great amount of data on solubilities and thermodynamics of solution of inert gases in various liquids. This topic has been reviewed extensively by various authors.[4,20-24] Here we shall present only a small sample of data on the solvation thermodynamics of inert gases in organic liquids. One of the main purposes of this presentation is to emphasize the sharp contrast between solvation of a simple molecule, such as argon, in an organic liquid and in water (Section 2.9).

It is worthwhile noting some general trends in the variations of these quantities that are revealed in Table 2.8. In all cases, values of ΔG_s^* are relatively large and positive for the small solvaton. As the solvaton increases from He to Xe, ΔG_s^* becomes smaller and even changes sign. This trend is probably due to the competition between the work required to form a cavity in the liquid, which probably gives a positive contribution to ΔG_s^*, and the soft interaction between the solvaton and its environment, which in this case is likely to give a negative contribution. As the inert gas molecule increases from He to Xe, it seems that the latter effect becomes the dominating one. The formal split of the solvation Gibbs energy into various contributions (cavity, dispersion forces, electrostatic interactions, etc.) is discussed in Section 3.5.

One further trend to be noted is that, for the small inert gas molecule, the solvation Gibbs energy is dominated by the $T\Delta S_s^*$ term, while as we increase the inert gas molecule there is a clearcut shift into enthalpy domination of the solvation Gibbs energy.

2.9. SOLVATION OF INERT GAS MOLECULES IN WATER

In this and the following section we present data on the solvation thermodynamics of simple solutes, such as inert gases or low hydrocarbons, in light and heavy water. These systems have been the target of extensive research ever since it was recognized that liquid water is an outstanding liquid, both in its pure state and as a solvent for various solutes.

Table 2.8
Solvation Thermodynamics of Inert Gases in Some Organic Liquids at 298.15 K[a]

	ΔG_s^*	ΔH_s^*	$T\Delta S_s^*$
n-Hexane			
He	7.506	− 6.39	− 13.90
Ne	6.635	− 8.79	− 15.43
Ar	1.887	− 17.13	− 19.02
Kr	− 0.640	− 19.15	− 18.51
Xe	− 3.895	− 25.12	− 21.23
n-Heptane			
He	7.891	− 6.56	− 14.40
Ne	7.017	− 8.67	− 15.69
Ar	2.192	− 15.46	− 17.66
Kr	− 0.414	− 19.76	− 19.34
n-Octane			
He	8.247	− 5.98	− 14.23
Ne	7.222	− 7.10	− 14.33
Ar	2.510	− 14.40	− 16.91
Kr	− 0.138	− 19.05	− 18.91
n-Nonane			
He	8.523	− 4.17	− 12.69
Ne	7.594	− 7.57	− 15.16
Ar	2.682	− 15.33	− 18.01
Kr	0.075	− 18.51	− 18.58
n-Decane			
He	8.791	− 5.92	− 14.72
Ne	7.803	− 7.26	− 15.06
Ar	2.874	− 15.15	− 18.03
Kr	0.263	− 18.55	− 18.81
n-Dodecane			
He	9.263	− 6.15	− 15.41
Ne	8.418	− 6.38	− 14.80
Ar	3.201	− 13.99	− 17.20
Kr	0.527	− 17.61	− 18.13
Xe	− 2.954	− 22.77	− 19.81
n-Tetradecane			
He	9.569	− 7.15	− 16.72
Ne	8.586	− 7.15	− 15.74
Ar	3.452	− 14.51	− 17.96
Kr	0.711	− 18.36	− 19.07
Cyclohexane			
He	7.661	− 6.31	− 13.97

Table 2.8 (*continued*)

	ΔG_s^*	ΔH_s^*	$T\Delta S_s^*$
Cyclohexane (*continued*)			
Ne	6.673	− 9.46	− 16.14
Ar	2.088	− 16.23	− 18.32
Kr	− 0.385	− 19.43	− 19.04
Xe	− 3.456	− 24.52	− 21.07
Benzene			
He	9.565	− 5.48	− 15.04
Ar	3.523	− 14.25	− 17.77
Kr	0.711	− 17.30	− 18.01
Xe	− 2.879	− 22.58	− 19.70
Toluene			
He	9.426	− 4.74	− 14.17
Ne	8.523	− 6.92	− 15.44
Ar	3.448	− 14.17	− 17.83
Kr	0.657	− 18.87	− 19.53
Xe	− 2.996	− 23.43	− 20.43
Fluorobenzene			
He	8.715	− 7.48	− 16.20
Ne	7.924	− 8.17	− 16.09
Ar	2.999	− 15.55	− 18.55
Kr	0.368	− 18.62	− 18.99
Xe	− 3.021	− 23.19	− 20.17
Chlorobenzene			
He	10.167	− 5.31	− 15.48
Ne	9.305	− 5.71	− 15.02
Ar	3.916	− 14.88	− 18.79
Kr	1.059	− 18.38	− 19.44
Xe	− 2.682	− 24.03	− 21.35
Bromobenzene			
He	10.874	− 4.12	− 15.00
Ne	9.920	− 6.94	− 16.86
Ar	4.577	− 14.38	− 18.96
Kr	1.582	− 18.36	− 19.94
Xe	− 2.280	− 22.86	− 20.58
Iodobenzene			
He	11.920	− 1.84	− 13.76
Ne	11.071	− 4.72	− 15.79
Ar	5.481	− 12.56	− 18.04
Kr	2.477	− 16.97	− 19.45
Xe	1.627	− 22.21	− 20.58

[a] Values computed are based on data taken from Wilhelm and Battino.[22] All values are in kJ mol^{-1}.

We start in this section with the simplest solutes in pure water. Perhaps the earliest systematic study of these solutions was undertaken by Eley,[25,26] who noted that some thermodynamic quantities of solution of inert gases in water are outstandingly different from the corresponding behavior of the same solutes in simple organic liquids. Eley also attempted to interpret his observation on a molecular level, basing his arguments on lattice models for water (a fashionable approach, at that time, to the theory of the liquid state).

Before describing the various attempts to interpret the outstanding features, or anomalies, of these solutions, we present a sample of experimental data to demonstrate the difference between aqueous and nonaqueous solutions of inert gas molecules. (More extensive reviews are available, e.g., Battino and Clever,[20] Wilhelm and Battino,[22] Ben-Naim,[4] and Franks.[24])

(1) The solubility, as measured by the Ostwald absorption coefficient, is markedly smaller in water as compared with a typical organic liquid. (By "typical" or "normal" organic liquid, we mean alkanes, alkanols, benzene and its simple derivatives, and so on.)

The Ostwald absorption coefficient is defined as the ratio of the number densities of a solute s in the liquid l and in the gaseous g phases at equilibrium, i.e.,

$$\gamma_s = (\rho_s^l/\rho_s^g)_{eq} \tag{2.35}$$

The solvation free energy of s in l is directly related to γ_s through

$$\Delta G_s^* = kT \ln(\rho_s^g/\rho_s^l)_{eq} = -kT \ln \gamma_s \tag{2.36}$$

Thus the statement of a low solubility, in terms of γ_s, is equivalent to a relatively large and positive value of the solvation free energy, say of argon in water, as compared with other organic solvents. In Figure 2.12 we plot the solvation Gibbs energies of argon and xenon in a series of alkanols as a function of n, the number of carbon atoms in the n-alkanol molecule (with $n = 0$ corresponding to liquid water). We see that the values of ΔG_s^* change very little as a function of n. If we attempt to extrapolate the value of ΔG_s^* for water by taking the limit $n \rightarrow 0$, we find that these values (indicated by arrows in Figure 2.12) are much lower than the experimental values. Thus a distinctly abrupt change in ΔG_s^* is observed when we pass from the series of n-alkanols to liquid water. In other words, water, though formally belonging to this homologous series with $n = 0$, has outstanding properties and cannot be viewed as one of the members of the homologue series of n-alkanols. A more detailed behavior of ΔG_s^* of inert gases in water as a function of temprature is shown in Figure 2.13 (based on data from Crovetto et al.[27]).

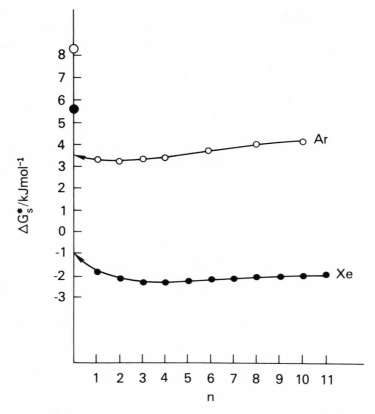

Figure 2.12. Solvation Gibbs energies of argon and xenon in n-alkanols as a function of n, the number of carbon atoms in the n-alkanol. The corresponding values for water are indicated at $n = 0$. All values correspond to 25 °C and 1 atm.

(2) The second striking difference between aqueous and nonaqueous solutions is the temperature dependence of the solvation Gibbs energy. This is manifested in the relatively large entropy of solvation of the inert gases in water as compared with other liquids. In Figure 2.14 we plotted values of ΔH_s^* and $T\Delta S_s^*$ for xenon in a series of n-alkanols. Again, we see that there is a very small variation in these values as we change n (the variation between the extreme values is believed to lie within the range of experimental error). The corresponding values of ΔH_s^* and $T\Delta S_s^*$ in water are markedly lower than the values that one would have extrapolated for the case of $n = 0$. A more detailed behavior of ΔS_s^* and ΔH_s^* as a function of T is shown in Figures 2.15 and 2.16 (based on data from Crovetto et al.[27]).

It is noteworthy that although both ΔH_s^* and ΔS_s^* are derived from the temperature dependence of ΔG_s^*, they reflect distinctly different aspects of the solvation process. In thermodynamic terms, we have

$$\Delta S_s^* = - (\partial \Delta G_s^*/\partial T)_P \tag{2.37}$$

$$\Delta H_s^* = \Delta G_s^* + T\Delta S_s^* = kT^2(\partial \ln \gamma_s/\partial T)_P \tag{2.38}$$

Hence ΔS_s^* measures directly the temperature coefficient of the solvation Gibbs energy, while ΔH_s^* is a measure of the temperature coefficient of γ_s or the concentration ratio of s between the two phases. For a molecular interpretation of these two quantities, see Section 3.8.

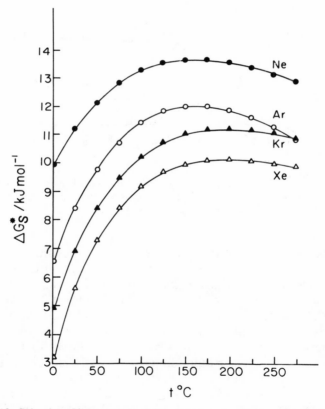

Figure 2.13. Solvation Gibbs energies of inert gas molecules in water as a function of temperature.

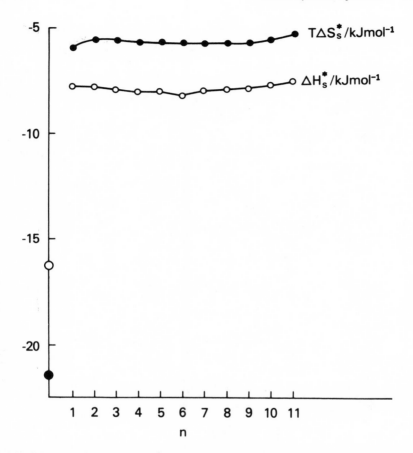

Figure 2.14. Values of $T\Delta S_s^*/\text{kJ mol}^{-1}$ and $\Delta H_s^*/\text{kJ mol}^{-1}$ for xenon in water ($n = 0$) and in a series of n-alkanols (n being the number of carbon atoms of the n-alkanols). All values pertain to 20 °C and 1 atm.

(3) The third, quite outstanding property of aqueous solutions of inert gas molecules is the relatively large partial molar heat capacity of the solute in water. This is equivalent to a large heat capacity of solvation of the inert gases in water.

If we define the heat capacity of solvation as

$$\Delta C_{p,s}^* = \partial \Delta H_s^* / \partial T \tag{2.39}$$

we find that for inert gas molecules in typical organic liquids, $\Delta C_{p,s}^*$ are within the experimental error, nearly zero. On the other hand, at about

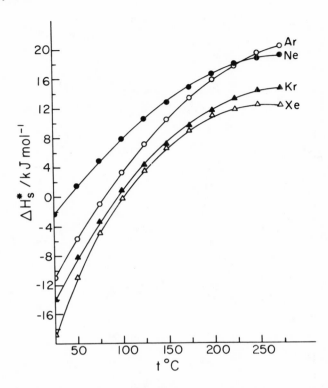

Figure 2.15. Values of $\Delta H_s^*/\text{kJ mol}^{-1}$ for Ne, Ar, Kr, and Xe in water as a function of temperature.

room temperature ΔH_s^* has a clearcut positive slope as a function of temperature, which gives rise to values of $\Delta C_{p,s}^*$ in the range of 150 to 250 $\text{J K}^{-1} \text{mol}^{-1}$ for the inert gas molecules in water at about room temperature (see Figure 2.17, based on data from Crovetto *et al.*[27]).

(4) Finally, we mention the volume of solvation, which is defined as

$$\Delta V_s^* = (\partial \Delta G_s^*/\partial P)_T \tag{2.40}$$

These may be computed from partial molar volumes of the inert gases in water and in other typical organic liquids. Table 2.9 shows some values of ΔV_s^* for methane and ethane in some liquids at 25 °C. These values are somewhat smaller in water as compared with the other liquids for which the relevant data are available. The differences are, however, not very large as we have noted for the other quantities of solvation.

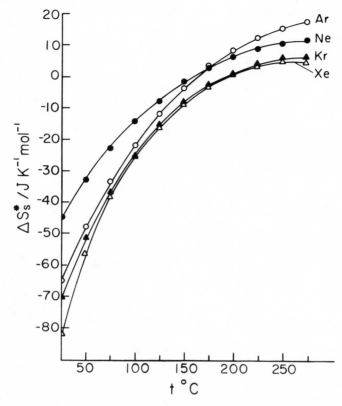

Figure 2.16. Values of $\Delta S_s^*/(J\ K^{-1}\ mol^{-1})$ for Ne, Ar, Kr, and Xe in water as a function of temperature.

Table 2.9
Solvation Volume $\Delta V_s^*/cm^3\ mol^{-1}$
for Methane and Ethane at 25 °C at
Infinite Dilution in Some Solvents

Solvent	Methane	Ethane
Water	36.17	50.07
Carbon tetrachloride	49.06	63.36
n-Hexane	56.03	65.33
Benzene	54.61	70.61

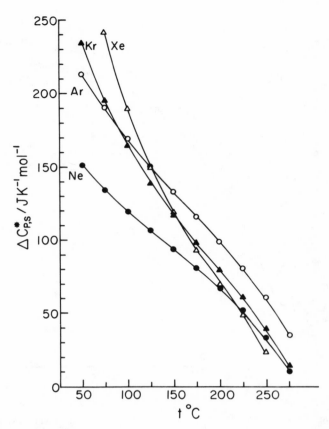

Figure 2.17. Values of $\Delta C_{p,s}^*/(J\ K^{-1}\ mol^{-1})$ for Ne, Ar, Kr, and Xe in water as a function of temperature.

Presently, there exists no theory of aqueous solutions that offers a satisfactory explanation of the outstanding properties of these systems. There are, however, some partial interpretations to some of the properties. Here we shall mention, only briefly, two approaches: both of them view the solvation process as comprising two steps. The first, and older one, splits the process of solvation (or rather the process of solution in the original treatment) into a cavity formation and a soft interaction between the solute and solvent. The second is based on the recognition that liquid water has some degree of structure, and this structure might be affected by the process of solvation. Hence one splits the process of solvation into two steps: first the solute is introduced into the solvent in which the structure has been "frozen in," and then the structure is allowed to relax to its final equilibrium value in the presence of the solute.

Both approaches contain an element of ambiguity. In the first, it is the size of the cavity that is presumed to accommodate the solute. For real solutes and solvents, there is no clearcut way of defining the size of such a cavity. One usually assigns an effective HS diameter to the molecules involved and then defines the radius of the appropriate cavity as the arithmetic mean of the solute and solvent diameters (for more details, see also Section 3.5). Once the choice of the cavity radius has been made, one can proceed in a rigorous and formal way and split all the thermodynamic quantities of solvation into two contributions. This procedure is outlined in detail in Section 3.5.

The second approach requires a definition of the structure of water. The essential idea was advanced by Frank and Evans,[28] who postulated that the large and negative entropy and enthalpy of solution of inert gas molecules might be due to "iceberg" formation. This idea has evolved into various modelistic approaches to aqueous solutions (some of these have been reviewed elsewhere[4]). A prototype of such an approach based on one particular definition of the structure of water is presented in Section 3.8.

2.10. SOLVATION OF INERT GAS MOLECULES IN LIGHT AND HEAVY WATER

Perhaps one of the earliest measurements of the solvation thermodynamics of argon in H_2O, D_2O, and mixtures of the two was published by Ben-Naim.[29] The main motivation for undertaking those measurements was to test the validity of some structural models that were used at that time to interpret the anomalous behavior of aqueous solutions of inert gas molecules. Table 2.10 presents some of the data on inert gases in light and heavy water.

It was found that the solvation Gibbs energy of argon, methane, and ethane in D_2O is smaller than in H_2O and that the Gibbs energy of solvation of argon changes quite smoothly when the composition of H_2O–D_2O is changed. These results were unexpected at that time. In the first place the large and positive value of ΔG_s^* for argon in water was attributed to the high degree of structure of H_2O as compared to other liquids. It was therefore argued that, if we replace H_2O with D_2O, the latter being known to have a higher degree of structure compared to H_2O, we should find that ΔG_s^* would be larger in D_2O than in H_2O. The experimental finding to the contrary was only qualitatively explained recently,[30] basing the argument on a specific definition of the structure of

Table 2.10
Values of ΔG_s^*/kJ mol^{-1}, ΔS_s^*/JK^{-1} mol^{-1}, ΔH/kJ mol^{-1},
and $\Delta C_{p,s}^*$/JK^{-1} mol^{-1} for
Neon, Argon, Krypton, and Xenon in H_2O and D_2O^a

t (°C)		ΔG_s^*	ΔS_s^*	ΔH_s^*	$\Delta C_{p,s}^*$
Neon					
25	(H_2O)	11.19	− 45.4	− 2.3	
	(D_2O)	10.99	− 43.6	− 2.0	
50	(H_2O)	12.14	− 33.2	1.4	150
	(D_2O)	11.90	− 32.2	1.5	140
75	(H_2O)	12.81	− 23.1	4.8	130
	(D_2O)	12.56	− 22.9	4.6	120
100	(H_2O)	13.26	− 14.7	7.8	120
	(D_2O)	13.02	− 15.2	7.3	110
Argon					
25	(H_2O)	8.40	− 65.1	− 11.0	
	(D_2O)	8.15	− 68.5	− 12.26	
50	(H_2O)	9.76	− 47.8	− 5.7	210
	(D_2O)	9.60	− 50.8	− 6.8	220
75	(H_2O)	10.74	− 33.6	− 0.9	190
	(D_2O)	10.64	− 36.0	− 1.9	200
100	(H_2O)	11.39	− 21.8	3.3	170
	(D_2O)	11.34	− 23.5	2.6	180
Krypton					
25	(H_2O)	6.94	− 70.5	− 14.1	
	(D_2O)	6.99	− 68.4	− 13.4	
50	(H_2O)	8.40	− 51.5	− 8.2	230
	(D_2O)	8.43	− 50.8	− 8.0	220
75	(H_2O)	9.46	− 36.8	− 3.3	200
	(D_2O)	9.48	− 36.7	− 3.3	190
100	(H_2O)	10.21	− 25.4	0.7	160
	(D_2O)	10.23	− 25.4	0.75	160
Xenon					
25	(H_2O)	5.62 '	− 81.9	− 18.8	
	(D_2O)	5.64	− 70.9	− 15.5	
50	(H_2O)	7.28	− 56.7	− 11.0	310
	(D_2O)	7.18	− 55.2	− 10.7	200
75	(H_2O)	8.41	− 38.6	− 5.0	240
	(D_2O)	8.34	− 41.0	− 5.9	190
100	(H_2O)	9.17	− 25.4	− 0.3	190
	(D_2O)	9.17	− 28.3	− 1.4	180

^{a} Based on data from Crovetto et al.[27]

water. Similarly, the fact that ΔG_s^* changed gradually upon adding D_2O to H_2O indicated that the structure of the mixture changes continuously from pure H_2O to pure D_2O, which means that D_2O molecules can replace H_2O molecules, not only without disrupting the structure of the liquid, but in fact reinforcing it in a gradual manner. Recently, the isotope effect on the Gibbs energy of solvation of ideal gases has been interpreted in terms of the changes induced by the solvation process on the structure of water.[31] Some details on this theory are presented in Section 3.9. The result is the following:

$$\Delta G_s^*(D_2O) - \Delta G_s^*(H_2O) = (\langle HB \rangle_s - \langle HB \rangle_0)(\varepsilon_D - \varepsilon_H) \qquad (2.41)$$

where on the lhs we have the difference in the solvation Gibbs energies of s in D_2O and H_2O. On the rhs we have the quantity $\langle HB \rangle_s - \langle HB \rangle_0$, which is the change in the average number of hydrogen bonds in the system induced by placing a solute s at a fixed position. The quantity $\varepsilon_D - \varepsilon_H$ is the difference in the hydrogen bond energies for D_2O and H_2O. Therefore from equation (2.41), one can estimate the change in the average number of hydrogen bonds induced by the solvation process. Some representative values of $\langle HB \rangle_s - \langle HB \rangle_0$ are given in Table 2.11. More details on these results can be found in a recent paper by Marcus and Ben-Naim.[32]

Table 2.11
Some Values of $\langle HB \rangle_s - \langle HB \rangle_0$ Calculated from Equation (2.41) Based on the Choice
$\varepsilon_D - \varepsilon_H = -0.929$ kJ mol^{-1} [a]

	t (°C)	$\langle HB \rangle_s - \langle HB \rangle_0$
Argon[29]	5	0.29
	10	0.26
	15	0.24
	20	0.22
	25	0.21
Neon[27]	25	0.22
Argon[27]	25	0.26
Krypton[27]	25	−0.06
Xenon[27]	25	−0.02
Methane[27]	25	0.08

[a] For more details, see Marcus and Ben-Naim.[32]

2.11. SOLVATION OF WATER IN PURE WATER

In Section 2.4 we studied one example of the solvation of a molecule in its pure liquid. Here, we apply the same concept to liquid water, perhaps the most studied liquid in physical chemistry. In the past the unique structural features of liquid water were studied through the solvation thermodynamics of various solutes in very dilute aqueous solutions. Here, on the other hand, we follow the solvation thermodynamics of a water molecule itself, a study never before undertaken.

The raw experimental data employed for the present calculations are densities of the liquid and the vapor pressure of water at various temperatures. It is assumed that the vapor at equilibrium with water at most temperatures of interest may be viewed as being an ideal gas. Of course, at very high temperatures, when we approach the critical point, we must use the densities rather than vapor pressures of the gaseous phase to compute the difference in the solvation properties in the liquid and gaseous phases.

Figure 2.18 shows values of the solvation Gibbs energy of H_2O, D_2O, and T_2O. Note that the values of ΔG_s^* for these molecules are considera-

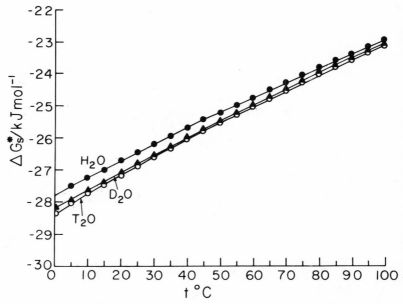

Figure 2.18. Solvation Gibbs energies of H_2O, D_2O, and T_2O in their respective pure liquids as a function of temperature.

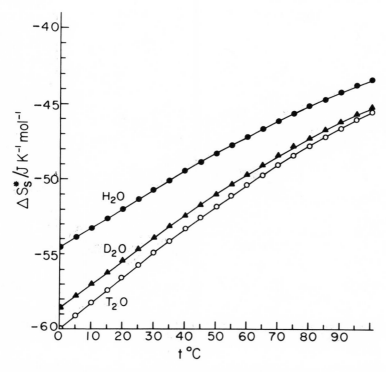

Figure 2.19. Solvation entropies of H_2O, D_2O, and T_2O in their respective pure liquids as a function of temperature.

bly more negative than the corresponding values of the inert gas molecules in water (Section 2.10). Figures 2.19 through 2.21 show also the values of ΔS_s^*, ΔH_s^*, and $\Delta C_{p,s}^*$ for the solvation of the various molecules in these liquids. Table 2.12 also presents some numerical values of the solvation thermodynamics of H_2O, D_2O, and T_2O in their own pure liquids. Further details on the method of computation of these quantities may be found elsewhere.[32] We note that in contrast to the case of the solvation of inert gas molecules in water, here the enthalpy rather than the entropy of solvation is the dominating part in ΔG_w^*. Thus the large negative value of ΔG_w^* of a water molecule in these liquids is primarily determined by the large solvation energy. This in turn reflects the strong binding energy of the molecule to its environment (see Section 3.8).

Values of $|\Delta H_w^*|$ decrease almost linearly as we increase the temperature. This presumably reflects the breakdown of some of the hydrogen bonds in the liquid, hence weakening the average binding energy of a

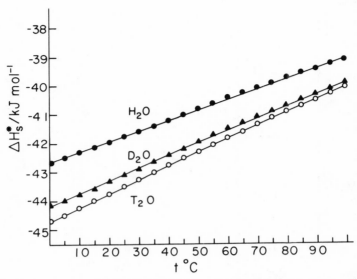

Figure 2.20. Solvation enthalpies of H_2O, D_2O, and T_2O in their respective pure liquids as a function of temperature.

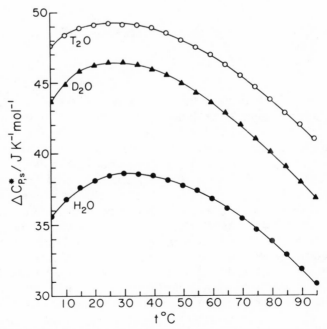

Figure 2.21. Solvation heat capacities (at constant pressure) of H_2O, D_2O, and T_2O in their respective pure liquids as a function of temperature.

Table 2.12

Solvation Thermodynamical Quantities of H_2O, D_2O, and T_2O as a Function of Temperature. The Pressure Corresponds to the Vapor Pressure at Equilibrium with the Liquid at Each Temperature[a]

	Temperature (°C)	ΔG_w^* (kJ mol^{-1})	ΔS_w^* (JK^{-1} mol^{-1})	ΔH_w^* (kJ mol^{-1})	$\Delta C_{p,w}^*$ (JK^{-1} mol^{-1})
H_2O	0	−27.792	−54.56	−42.695	
H_2O	5	−27.521	−53.92	−42.52	35.4
D_2O		−27.94	−57.8	−44.018	43.3
T_2O		−28.042	−59.1	−44.48	47.4
H_2O	10	−27.253	−53.28	−42.338	36.6
D_2O		−27.653	−57.01	−43.797	44.7
T_2O		−27.748	−58.24	−44.24	48.2
H_2O	15	−26.988	−52.62	−42.152	37.4
D_2O		−27.37	−56.22	−43.569	45.6
T_2O		−27.459	−57.39	−43.996	48.8
H_2O	20	−26.727	−51.97	−41.962	38.0
D_2O		−27.091	−55.43	−43.339	46.2
T_2O		−27.174	−56.55	−43.751	49.1
H_2O	25	−26.468	−51.33	−41.771	38.4
D_2O		−26.816	−54.64	−43.107	46.4
T_2O		−26.894	−55.72	−43.505	49.3
H_2O	30	−26.213	−50.68	−41.578	38.6
D_2O		−26.545	−53.87	−42.875	46.5
T_2O		−26.617	−54.9	−43.259	49.3
H_2O	35	−25.962	−50.05	−41.385	38.6
D_2O		−26.278	−53.11	−42.643	46.3
T_2O		−26.345	−54.09	−43.013	49.2
H_2O	40	−25.713	−49.43	−41.193	38.5
D_2O		−26.014	−52.37	−42.413	46.0
T_2O		−26.076	−53.3	−42.768	48.9
H_2O	45	−25.467	−48.83	−41.001	38.2
D_2O		−25.754	−51.64	−42.184	45.6
T_2O		−25.812	−52.53	−42.524	48.6
H_2O	50	−25.225	−48.23	−40.812	37.9
D_2O		−25.497	−50.94	−41.958	45.1
T_2O		−25.551	−51.78	−42.283	48.2
H_2O	55	−24.985	−47.66	−40.624	37.5
D_2O		−25.244	−50.25	−41.735	44.4
T_2O		−25.294	−51.04	−42.044	47.6
H_2O	60	−24.748	−47.1	−40.438	36.9
D_2O		−24.995	−49.59	−41.516	43.7
T_2O		−25.041	−50.33	−41.808	47.0

(continued)

Table 2.12 (*continued*)

	Temperature (°C)	ΔG_w^* (kJ mol^{-1})	ΔS_w^* (JK^{-1} mol^{-1})	ΔH_w^* (kJ mol^{-1})	$\Delta C_{p,w}^*$ (JK^{-1} mol^{-1})
H_2O	65	−24.514	−46.55	−40.256	36.3
D_2O		−24.749	−48.95	−41.3	42.9
T_2O		−24.791	−49.64	−41.575	46.4
H_2O	70	−24.283	−46.03	−40.077	35.6
D_2O		−24.505	−48.33	−41.088	42.1
T_2O		−24.544	−48.96	−41.346	45.6
H_2O	75	−24.054	−45.52	−39.902	34.8
D_2O		−24.265	−47.73	−40.881	41.2
T_2O		−24.301	−48.31	−41.121	44.8
H_2O	80	−23.827	−45.03	−39.731	34.0
D_2O		−24.028	−47.15	−40.679	40.2
T_2O		−24.061	−47.68	−40.9	44.0
H_2O	85	−23.603	−44.57	−39.564	33.1
D_2O		−23.794	−46.59	−40.481	39.2
T_2O		−23.824	−47.07	−40.684	43.1
H_2O	90	−23.382	−44.12	−39.403	32.1
D_2O		−23.562	−46.06	−40.289	38.2
T_2O		−23.59	−46.49	−40.472	42.1
H_2O	95	−23.162	−43.69	−39.246	31.1
D_2O		−23.333	−45.55	−40.102	37.1
T_2O		−23.359	−45.92	−40.265	41.1
H_2O	100	−22.945	−43.28	−39.095	30.0
D_2O		−23.107	−45.06	−39.921	36.0
T_2O		−23.131	−45.38	−40.063	40.1

[a] Data from Marcus and Ben-Naim.[32]

water molecule to its environment. Replacing H_2O by D_2O and by T_2O makes ΔH_w^* more negative, presumably a result of the strengthening of the average binding energy of a water molecule.

We observe a similar trend in the variation of the entropy as a function of temperature and upon replacing H_2O by D_2O and T_2O. The results seem to be consistent with the presently accepted fact that the heavier isotopes of water are more structured liquids.

We note also that the heat capacity of solvation of a water molecule has a maximum value between 25 and 30 °C for these liquids.

In Section 3.9 we describe a method of extracting information on the structure of water (as defined in Section 3.7) from the isotope effect on the solvation Gibbs energy of a water molecule. The result is

$$\Delta G^*_{D_2O}(D_2O) - \Delta G^*_{H_2O}(H_2O) = \tfrac{1}{2}\langle \psi_w \rangle (\varepsilon_D - \varepsilon_H) \qquad (2.42)$$

where on the lhs of equation (2.42) we have the experimental quantities of the solvation Gibbs energies of H_2O in H_2O and of D_2O in D_2O. With a choice of $\varepsilon_D - \varepsilon_H = -0.929$ kJ mol^{-1} we can evaluate $\langle \psi_w \rangle$, the average number of hydrogen bonds in which a water molecule participates at that temperature (for more details on this topic, see Marcus and Ben-Naim[32]). We present in this section one set of these results in Figure 2.22, where $\langle \psi_w \rangle$ is plotted as a function of temperature. This number decreases monotonically from about 0.95 at 0 °C to about 0.35 at 100 °C. Further assumptions[32] lead to an approximate value of the average structure of H_2O and D_2O separately. These are designated in Figure 2.22 as $\langle \psi_w \rangle^H$ and $\langle \psi_w \rangle^D$, respectively. It is clear from the figure that the structure of D_2O is somewhat larger than that of H_2O. These findings are consistent, based on a variety of other experimental means, with conclusions reached by other authors.

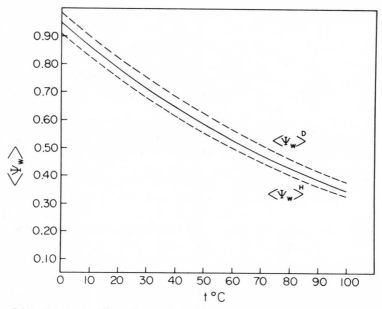

Figure 2.22. Average number of hydrogen bonds in which a water molecule participates as a function of temperature. $\langle \psi_w \rangle^H$ and $\langle \psi_w \rangle^D$ correspond to H_2O and D_2O (dashed curves). The full curve is the value of $\langle \psi_w \rangle$ at an intermediary point between H_2O and D_2O.

Of course, the values of $\langle \psi_w \rangle$ should not be taken too seriously. They have been obtained by too many approximations and assumptions (e.g., the form of the water–water pair potential, neglect of higher-order potentials, the estimate of $\varepsilon_D - \varepsilon_H$, and so on). Nevertheless, we believe that, within the approximations that were made, the general trend in the variation of $\langle \psi_w \rangle$, say as a function of temperature, gives a reasonable measure of the change of the structure of water.

2.12. LIMITING EFFECT OF SOLUTES ON THE SOLVATION THERMODYNAMICS OF A WATER MOLECULE

In Section 1.7 we presented the general relationship between the effect of a solute on the solvation of a water molecule and experimentally measurable quantities. In this section we present some numerical results for the limiting effect of various solutes on the solvation thermodynamics of a water molecule.

As noted earlier, the structural aspects of dilute aqueous solutions have been traditionally investigated through the solvation properties of the *solute* rather than the water molecule itself. However, if we are interested in the structure of water itself or in its changes upon the addition of various solutes, it is clearly more advisable to study directly the solvation of a water molecule itself rather than the solvation of a foreign molecule [athough the two quantities are connected through the Gibbs–Duhem relation (see Section 1.8)].

In Section 1.7 we introduced the quantities $\delta G^*(w/s)$, $\delta S^*(w/s)$, $\delta H^*(w/s)$, and so on, as the limiting effects of a solute s on the solvation quantities of a water moleclue.

Table 2.13 presents values of δG^* and some values of $T\delta S^*$ and δH^* for simple solutes. The values of δG^* were computed using data on partial molar volumes of the solutes at infinite dilution (see Section 1.7).[33] In most cases such data are available for one temperature only. In those few cases where the temperature dependence of the partial molar volume is also reported, we have calculated $T\delta S^*$ and δH^* by taking the average derivative between the two available temperatures. As can be seen from Table 2.13, all values of δG^* for the simple solutes in water are positive. In a homologous series of the hydrocarbons, δG^* is roughly linear in the number of carbon atoms in the molecule. Since ΔG_s^* of water in water are negative at these temperatures, the effect of the solutes on ΔG_s^* is to reduce its absolute magnitude. A glance at the values of $T\delta S^*$ and δH^* shows that this reduction comes about from the entropy rather than the enthalpy changes.

The large and negative values of $T\delta S^*$ are probably a result of the

Table 2.13
Values of δG^*, $T\delta S^*$, and δH^* (all in kJ/mol) for
Small Solutes in Water at Various Temperatures

Solute	T/K	δG^*	$T\delta S^*$	δH^*
Hydrogen	298.15	1.18		
Nitrogen	298.15	2.42		
Oxygen	298.15	2.07		
Argon	298.15	1.99		
Krypton	298.15	2.02		
Carbon monoxide	298.15	2.64		
Methane	289.95	2.03		
	296.15	2.48	− 17	− 14.8
	302.25	2.76		
Ethane	290.05	4.04		
	296.15	4.42	− 16	− 11.8
	302.25	4.72		
Propane	290.05	6.10		
	296.15	6.61	− 17	− 11.5
	302.25	6.86		
n-Butane	298.15	8.03		
Benzene	283.25	8.72		
	298.35	8.94		
	303.15	9.08	− 11.8	− 2.9
	308.05	9.31		
	313.05	9.51		

[a] Data from Ben-Naim.[33]

increase in the average structure of water around a water molecule induced by the addition of the solute. Similarly, the negative values of δH^* reflect the increase in the average binding energy of a water molecule to its environment caused by the addition of the solute.

We also noted in Section 1.7 a relation between δG^* and the relative affinities of water toward water or toward a solute molecule. From this point of view the positive values of δG^* indicate what is intuitively obvious — water–water affinity is greater than water–solute affinity. The difference between the two becomes greater the larger the solute molecule.

Figure 2.23 and Table 2.14 show the dependence of δG^* on the chain length of some homologous series of solute molecules. In all cases the

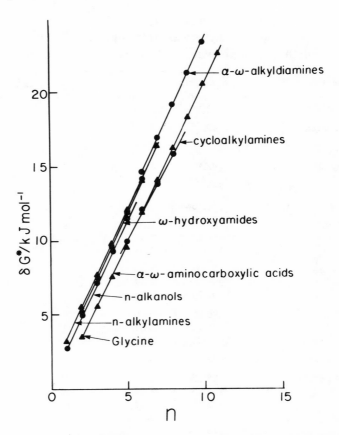

Figure 2.23. Dependence of $\delta G^*/kJ\ mol^{-1}$ on the number of carbon atoms in several homologous series of hydrocarbon molecules. (All values at 25 °C.)

linearity of the curves is quite remarkable. Furthermore, for a given chain length the value of δG^* increases when we replace an hydroxyl group by amide, and further increases upon replacement by an amine group. We also note that a similar trend is observed for the cyclohydrocarbons when replacing a ketone by hydroxyl and by amine groups. Also, replacing a single bond by a double or a triple bond causes a reduction in the values of δG^*.

Table 2.15 presents some values of δG^*, $T\delta S^*$, and δH^* for some ionic solutes in water at 25 °C. Here, the general trend is that δG^* changes from relatively large negative values to positive values as the ion size increases (some exceptions to this trend should be noted, however). This general trend is consistent with the interpretation of δG^* in terms of

Table 2.14
Values of δG^*(s/w) for Some Homologous Series of Hydrocarbons[a]

	δG^*/kJ mol^{-1}		δG^*/kJ mol^{-1}
n-Alkanols		Cycloamines	
Methanol	2.76	Cyclopentylamine	10.00
Ethanol	5.08	Cyclohexylamine	11.97
n-Propanol	7.20	Cycloheptylamine	13.77
n-Butanol	9.40	Cyclo-octylamine	15.82
n-Pentanol	11.56	1-w-Alkyldiamines	
n-Alkylamides		Ethylenediamine	6.11
Formamide	2.89	1,3-Propanediamine	8.23
Acetamide	5.21	1,4-Butanediamine	10.36
n-Propionamide	7.24	1,5-Pentanediamine	12.52
n-Butyramide	9.48	1,6-Hexanediamine	14.64
		1,7-Heptanediamine	16.93
n-Alkylamines		1,8-Octanediamine	19.09
Methyamine	3.24	1,9-Nonanediamine	21.22
Ethylamine	5.53	1,10-Decanediamine	23.35
n-Propylamine	7.68		
n-Butylamine	9.84	1-w-Alkyldiols	
n-Pentylamine	12.02	Ethylene glycol	5.12
n-Hexylamine	14.20	1,3-Propanediol	7.41
n-Heptylamine	16.39	1,4-Butanediol	9.66
		1,5-Pentanediol	11.81
Cycloketones		1,6-Hexanediol	14.06
Cyclobutanone	7.24	1,8-Octanediol	18.45
Cyclopentanone	9.11	1,10-Decanediol	22.81
Cyclohexanone	11.19		
Cycloheptanone	13.13	Ethane	4.78
Cyclo-octanone	15.07	Ethylene	4.56
Cyclononanone	16.99	Acetylene	3.35
		Propane	7.22
Cycloalkanols		Propylene	5.30
Cyclobutanol	7.89	n-Butane	8.03
Cyclopentanol	9.62	1-Butene	6.06
Cyclohexanol	11.61	1,3-Butadiene	6.89
Cycloheptanol	13.55	Cyclopropane	4.97
Cyclo-octanol	15.31	2-Methylpropane	8.92

[a] All values at 25 °C. Data from Ben-Naim.[33]

relative affinities (see Section 1.7), namely, as we increase the size of the ion the solute–water affinity decreases (toward values typical of nonionic solutes); hence, the quantity $G_{ww} - G_{ws}$ becomes more positive. This trend continues to be maintained when the ions become very large, as shown in Tables 2.16 and 2.17 for the case of tetraalkylammonium salts.

Table 2.15

Values of δG^*, $T\delta S^*$, and δH^* (all in kJ/mol) for Some Simple Electrolytes at 25 °C[a]

		F^-	Cl^-	Br^-	I^-
H^+	δG^*	-2.638	-0.037	0.910	2.488
	$T\delta S^*$	1.05	-1.92	-3.28	-6.30
	δH^*	-1.59	-1.95	-2.37	-3.81
Li^+	δG^*	-2.758	-0.154	0.789	2.367
	$T\delta S^*$	1.81	-1.19	-2.50	-5.54
	δH^*	-0.95	-1.35	-1.72	-3.17
Na^+	δG^*	-2.804	-0.199	0.823	2.322
	$T\delta S^*$	-2.24	-5.26	-7.51	-9.60
	δH^*	-5.05	-5.46	-6.70	-7.28
K^+	δG^*	-1.646	1.203	2.147	3.725
	$T\delta S^*$	-2.92	-5.93	-7.25	-9.08
	δH^*	-4.56	-4.72	-5.10	-5.35
Rb^+	δG^*	-0.708	1.896	2.839	4.418
	$T\delta S^*$	-3.46	-6.46	-7.78	-10.82
	δH^*	-4.16	-4.57	-4.94	-6.40

[a] Data from Ben-Naim.[33]

Table 2.16

Values of $\delta G^*/\text{kJ mol}^{-1}$ for Tetraalkylammonium Salts at 25 °C[a]

	Cl^-	Br^-	I^-
Me_4N^+	12.23	13.20	14.76
Et_4N^+	20.36	21.34	22.92
Pr_4N^+	29.37	30.34	31.91
Bu_4N^+	37.76	38.72	40.30
Pen_4N^+	—	47.36	—

[a] Data from Ben-Naim.[33]

Also worth noting are the general trends in the variation of δH^* and $T\delta S^*$ with ionic size. For very small ions δS^* are positive, indicating a net decrease in the structure around the solvated water molecule. As we increase the size of the ion, δS^* changes sign and becomes negative, as in the case of nonionic solutes. Similar trends are observed in the variation of δH^*. Note also that except for very small ions, the values of $|T\delta S^*|$ are larger than $|\delta H^*|$, i.e., the entropy part dominates the values of δG^*.

Table 2.17
Values of δG^*, $T\delta S$, and δH^* (all in kJ/mol) for Some Tetraalkylammonium Salts[a]

	T/K	δG^*	$T\delta S^*$	δH^*
NH_4Cl	273.15	2.004	− 5.15	− 3.15
	298.15	2.475	− 3.79	− 1.31
	323.15	2.793		
Me_4NCl	273.15	10.97	− 13.98	− 3.01
	298.15	12.25	− 13.74	− 1.46
	323.15	13.40		
Et_4NCl	273.15	18.39	− 22.18	− 3.79
	298.15	20.42	− 22.30	− 1.88
	323.15	22.29		
$n\text{-}Pr_4NCl$	273.15	26.59	− 30.37	− 3.78
	298.15	29.37	− 33.75	− 4.38
	323.15	32.20		
$n\text{-}Bu_4NCl$	273.15	33.97	− 41.52	− 7.55
	298.15	37.77	− 49.73	− 11.96
	323.15	41.94		

[a] Data from Ben-Naim.[33]

Table 2.18
Values of $\delta G^*/(kJ/mol)$ in H_2O and D_2O for Some Ionic Solutes at 25 °C[a]

	$\delta G^*(H_2O/S)$	$\delta G^*(D_2O/S)$
NaF	− 2.804	− 2.997
NaCl	− 0.200	− 0.322
NaBr	0.7631	0.706
KCl	1.196	1.134
KBr	2.139	2.098
Me_4NBr	13.191	13.124
Et_4NBr	21.34	21.293
$n\text{-}Pr_4NBr$	30.352	30.304
$n\text{-}Bu_4NBr$	38.795	38.791

[a] Data from Ben-Naim.[33]

Finally, we present in Table 2.18 values of δG^* for some salts in H_2O and D_2O.[33] It is seen that the difference $\delta G^*(D_2O/s) - \delta G^*(H_2O/s)$ is always negative. We may interpret these results as indicating that the s–H_2O affinity is stronger relative to the s–D_2O affinity (Section 1.7).

In order to obtain a more complete study of the effect of various solutes on the solvation thermodynamics of a water molecule, we need more accurate data on partial molar volumes of these solutes as well as their dependence on temperature and pressure. We believe that such a systematic study will be highly rewarding in illuminating the effect of solutes on both the structure and energetics of solvation phenomena of a water molecule in aqueous solutions.

2.13. SOLVATION THERMODYNAMICS OF WATER IN CONCENTRATED AQUEOUS SOLUTIONS

In Section 2.12 we examined the limiting effect of various solutes on the thermodynamics of solvation of a water molecule. This section extends the previous one into the more concentrated solutions. The CP of water in the liquid solution l is written in the conventional and statistical mechanical forms as follows:

$$\mu_w = \mu_w^{*l} + kT \ln \rho_w^l \Lambda_w^3$$
$$= \mu_w^p + kT \ln a_w = \mu_w^{*p} + kT \ln \rho_w^p \Lambda_w^3 a_w \qquad (2.43)$$

where μ_w^p is the CP of pure water and a_w is the activity of water in the solution. From equation (2.43) we may obtain the difference in the solvation Gibbs energy of water in the solution l relative to pure water, i.e.,

$$\Delta\Delta G_w^* = \Delta G_w^{*l} - \Delta G_w^{*p} = \mu_w^{*l} - \mu_w^{*p} = kT \ln(\rho_w^p a_w / \rho_w^l) \qquad (2.44)$$

Thus in order to evaluate $\Delta\Delta G_w^*$, we need the densities of water in pure water (ρ_w^p) and in the solution (ρ_w^l), as well as the activity of the water in the solution. The latter is usually given in terms of the molal osmotic coefficient ϕ_w defined by [34]

$$kT \ln a_w = -\nu m_s M_w kT \phi_w / 1000 \qquad (2.45)$$

where m_s is the molality of the solute, ν the number of ions produced by the dissociation of the solute (here we examine only 1:1 electrolytes for which $\nu = 2$), and M_w is the molecular mass of the water. Thus using available data on the osmotic coefficients of various electrolytes in water and the densities of water, we can compute values of $\Delta\Delta G_w^*$ as defined in relation (2.44).[33]

Figures 2.24, 2.25, and 2.26 show the values of $\Delta\Delta G_w^*$ for water, as a function of the molality of the solutes, at one temperature, 25 °C. For HCl

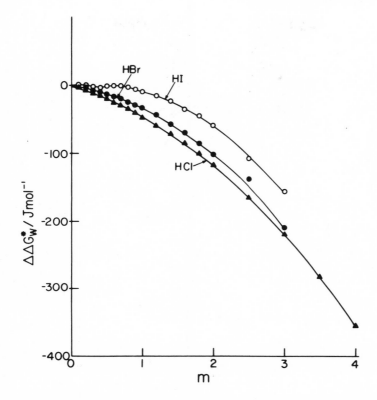

Figure 2.24. Values of $\Delta\Delta G_w^*$ for water in aqueous solutions of HCl, HBr, and HI as a function of the molality of the acids at 25 °C.

solutions all values of $\Delta\Delta G_w^*$ are negative. Since ΔG_w^{*p} (i.e., for water in pure water) at 25 °C is about -26.46 kJ mol^{-1}, we conclude that addition of HCl makes the solvation Gibbs energy of water in the solution more negative than in pure water. It is not clear whether this effect comes about from the *direct* solute–water interaction or through the *indirect* effect of the solute on the structure of the surroundings of the solvated water molecule. A qualitative reasoning leads to favoring the former effect. At any given solute concentration, if we replace Cl$^-$ by Br$^-$ and I$^-$, we find that $\Delta\Delta G_w^*$ becomes less negative (or even positive). This is probably due to weakening of the electrostatic interaction between the ions and the water solvaton. The trend is similar for the series NaCl, NaBr, and NaI, and even more pronounced for the series KCl, KBr, and KI.

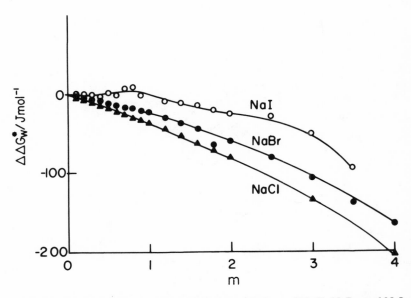

Figure 2.25. Values of $\Delta\Delta G_w^*$ for water in aqueous solutions of NaCl, NaBr, and NaI as a function of the molality of the acids at 25 °C.

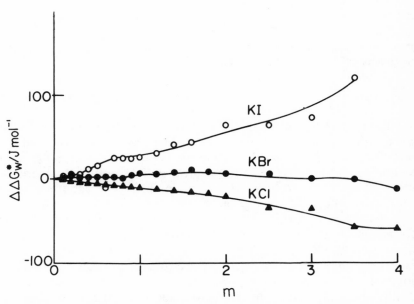

Figure 2.26. Values of $\Delta\Delta G_w^*$ for water in aqueous solutions of KCl, KBr, and KI as a function of the molality of the acids at 25 °C.

Regarding the entropy and enthalpy of solvation, the required experimental data are quite scarce. In Figure 2.27 we show the dependence of $\Delta\Delta G_w^*$ on the temperature for some concentrations of NaCl.[34] Assuming that these plots are nearly linear in T, we have computed some values of $\Delta\Delta S_w^*$ and $\Delta\Delta H_w^*$ which are reported in Table 2.19. Values of

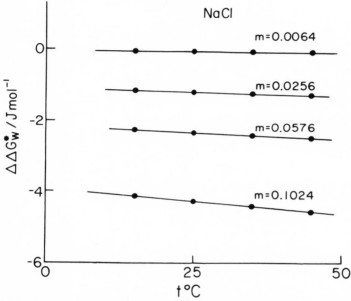

Figure 2.27. Temperature dependence of $\Delta\Delta G_w^*$ for some concentrations (m = molality) of NaCl.

Table 2.19
Values of $\Delta\Delta G_w^*$, $\Delta\Delta S_w^*$, and $\Delta\Delta H_w^*$ for Water in Some Aqueous Solutions of NaCl and KCl at 25 °C[a]

m	$\Delta\Delta G_w^*/\text{J mol}^{-1}$	$\Delta\Delta S_w^*/\text{J mol}^{-1}\,\text{K}^{-1}$	$\Delta\Delta H_w^*/\text{J mol}^{-1}$
		NaCl	
0.064	− 0.0609	1.81×10^{-4}	$- 6.93 \times 10^{-3}$
0.0256	− 1.1918	3.87×10^{-3}	$- 3.79 \times 10^{-2}$
0.0576	− 2.3655	7.675×10^{-3}	$- 7.72 \times 10^{-2}$
0.1024	− 4.3135	1.4135×10^{-2}	$- 9.91 \times 10^{-2}$
		KCl	
0.0064	− 0.3055	1.01×10^{-3}	$- 4.37 \times 10^{-3}$
0.0256	− 0.6928	2.185×10^{-3}	$- 4.13 \times 10^{-2}$
0.0576	− 1.3581	4.215×10^{-3}	$- 1.01 \times 10^{-1}$
0.1024	− 2.036	6.315×10^{-3}	$- 1.53 \times 10^{-1}$

[a] m is the molality of the salt. Data from Ben-Naim.[33]

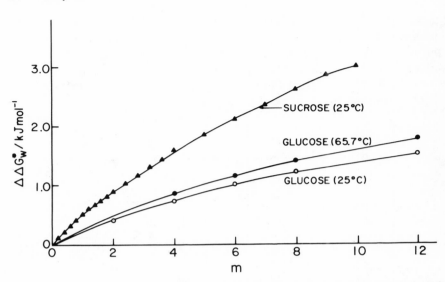

Figure 2.28. Values of $\Delta\Delta G_w^*$ for aqueous solutions of sucrose and glucose as a function of the molality of the solute.

$\Delta\Delta S_w^*$ are all positive; this might indicate a decrease in the structure of the environment of a solvated water molecule. On the other hand, all values of $\Delta\Delta H_w^*$ are negative, probably a result of the additional water-ion interaction, which tends to increase the overall strength of the binding energy of a water molecule to its surroundings.

For comparison, we have also calculated the solvation Gibbs energies of water in aqueous solutions of sucrose and glucose. The results are shown in Figure 2.28. Here, in contrast to the ionic solutes, we find positive values of $\Delta\Delta G_w^*$. The sign is the same as in the case of addition of ethanol (see Section 2.14). Again, it is not possible to state categorically whether this effect is due to the *direct* interaction between the solute and the solvated water molecule or from the *indirect* change of the structure of the environment of the solvated molecule induced by the addition of the solute.

2.14. SOLVATION IN WATER–ETHANOL MIXTURES

In Section 2.5 we discussed the solvation thermodynamics of argon and krypton in mixtures of argon and krypton. Here, we use the same method to study the solvation of both water and ethanol in mixtures of these two liquids. The raw data used for our calculation are excess Gibbs energies and excess volumes. However, since the data collected from various sources[35-38] do not correspond to the same temperature we have adopted the following approximate procedure.

The excess Gibbs energies are available at $T = 303.15$ K. On the other hand, the excess volumes are given at $T = 298.15$ K and $T = 308.15$ K. Therefore, we have estimated the value of $kT \ln(v_m/V_w^p)$ (see Section 1.5) at $T = 303.15$ K as an average between 298.15 K and 308.15 K. In this way approximate values of $\Delta\Delta G_w^*$ and $\Delta\Delta G_{Et}^*$ were calculated for both water and ethanol molecules, respectively, according to equation (1.87). We note that $\Delta\Delta G_w^*$ is the relative solvation Gibbs energy of a water molecule in a given mixture relative to the solvation Gibbs energy of water in pure water. Similarly $\Delta\Delta G_{Et}^*$ is the relative quantity with respect to the solvation Gibbs energy in pure ethanol.

The computed results are shown in Figures 2.29 and 2.30 for two temperatures, $T = 303.15$ K and $T = 323.15$ K. At both temperatures we see that $\Delta\Delta G_w^*$ are positive. Since ΔG_w^{*p} is negative, we conclude that the addition of ethanol makes the solvation Gibbs energy of a water molecule less negative. The major reason for this result is probably due to the

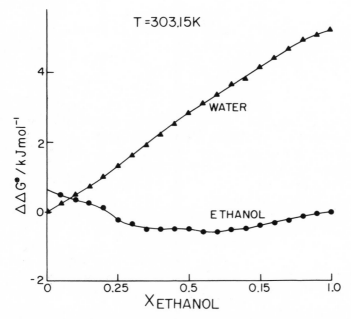

Figure 2.29. Values of $\Delta\Delta G^*$ for water and for ethanol as a function of the mole fraction of ethanol at $T = 303.15$ K.

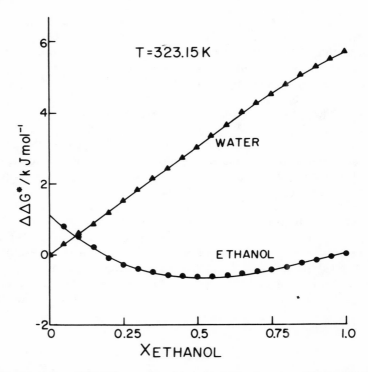

Figure 2.30. Values of $\Delta\Delta G^*$ for water and for ethanol as a function of the mole fraction of ethanol at $T = 323.15$ K.

overall weakening of the average binding energy of a water molecule to its surroundings when ethanol is added to the system. We expected that at very low ethanol concentration an opposite effect would be observed, since there are many indications that small amounts of ethanol added to water have a stabilizing effect on the structure of water. We do not know if this monotonic behavior of ΔG_w^* as a function of X_{Et} is a real phenomenon or an artifact due to the approximation used in our calculation.

On the other hand, adding water to pure ethanol makes ΔG_{Et}^* more negative than in pure ethanol (this is true except for very high concentrations of water where $\Delta\Delta G_{Et}^*$ become slightly positive). The interpretation is not clear, but this might be due to an overall strengthening of the total binding energy of ethanol to its surroundings, caused by the addition of water.

It would have been of interest to compute also $\Delta\Delta S^*$ and $\Delta\Delta H^*$ in this system. However, because of the inaccuracy of the whole procedure, the values that we can obtain from the available data are highly unreliable. We believe that the solvation thermodynamics of both water and ethanol in these systems is of sufficient interest to warrant more complete and accurate experimental measurements of excess function in these mixtures. In particular very accurate data are needed in the region of very low alcohol concentrations, and on the temperature dependence of all the relevant excess functions.

2.15. SOLVATION OF VARIOUS HYDROCARBON MOLECULES IN THEIR OWN PURE LIQUIDS

In this section we discuss the solvation thermodynamics of various hydrocarbons in their own pure liquids. Our starting relation is equation (2.9), where we have assumed that the vapor pressure of the liquids, around room temperature, is low enough so that the gaseous phase may be treated as an ideal gas. Hence we write

$$\Delta G_s^{*p} = kT \ln(P_s/kT\rho_s^l)_{eq} \qquad (2.46)$$

where P_s is the vapor pressure and ρ_s^l the density of the liquid s.

Figure 2.31 shows values of ΔG_s^* of some homologous series of hydrocarbons as functions of the number of carbon atoms in the alkyl (or alkene) chain. For details of the computation procedure and sources of experimental data, see Ben-Naim and Marcus.[39] As is discernible from this figure, the solvation Gibbs energies of these homologous series are almost linear functions of n. A qualitative interpretation of these findings may be given as follows.

Let us denote the members of the homologous series by $X-(CH_2)_{n-1}-CH_3$, where X may be a carboxylic group, benzyl group, etc. The linearity of ΔG_n^* with respect to n means that, to a good approximation, the increment in the solvation Gibbs energy $\Delta G_{n+1}^* - \Delta G_n^*$ is almost constant for a given homologous series (in fact, from Figure 2.31 it is clear that the increments are also almost independent of the functional group X, i.e., all the curves are nearly parallel to each other).

Now, in a given "solvent" l, we could write for any homologous series the equality[4]

$$\delta G(CH_3 + n \rightarrow n + 1) = \Delta G_{n+1}^{*l} - \Delta G_n^{*l} - \Delta G_{CH_3}^{*l} \qquad (2.47)$$

where δG is the indirect part of the Gibbs energy change for the process of bringing a methyl group and adding it to a chain of size n to form a new

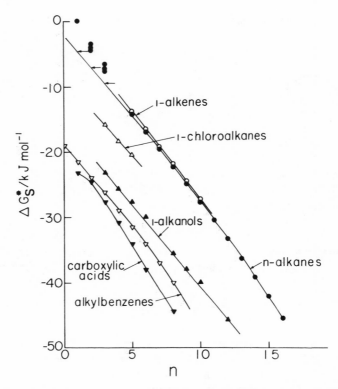

Figure 2.31. Gibbs energy of solvation of hydrocarbon molecules in their own pure liquids as a function of n, the number of carbon atoms in the alkyl chains. All values pertain to $T = 298.15$ K.

chain of size $n + 1$ (see also Section 3.18 on hydrophobic interaction, where this quantity is discussed in more detail). Of course, the data in Figure 2.31 do not apply to the *same* "solvent" l, but each molecule is solvated in its own pure liquid. We may assume, however, that for each pair of solvents n and $n + 1$, we can apply equation (2.47) to an intermediary solvent with properties between n and $n + 1$. We can thus arrive at the approximate conclusion that the indirect work of adding a methyl group to a chain of size n is almost independent of n. We shall see in Section 2.16 that a similar conclusion may be reached for hydrocarbons in liquid water (in this case the solvent, water, is common to all the hydrocarbons).

   In Figure 2.31 we have also marked three arrows at the points, on the n-alkane line, corresponding to the values extrapolated for methane,

ethane, and propane. We have also computed the solvation Gibbs energies of these small alkanes in some liquid n-alkanes using solubility data from the literature. All the ΔG_s^* values thus computed fall above the corresponding arrows; the farther away from the arrow, the larger the size of the solvent n-alkane in which ΔG_s^* has been computed. This observation can be interpreted in terms of the strength of the average binding energy of the solute to the solvent. As an example let us assume that the point of the arrow for ethane represents the solvation Gibbs energy of an ethane molecule in a hypothetical liquid ethane at room temperature (i.e. for $n = 2$ and roughly one atmospheric pressure). In such a case the solvaton will interact with only *methyl* groups of the surrounding molecules.

Now, we replace the "solvent" ethane by a real solvent, say n-hexane. Here the environment of the solvaton will consist of methyl and methylene groups. The overall binding energy of the solvaton to the solvent is expected to decrease (the argument being essentially the same

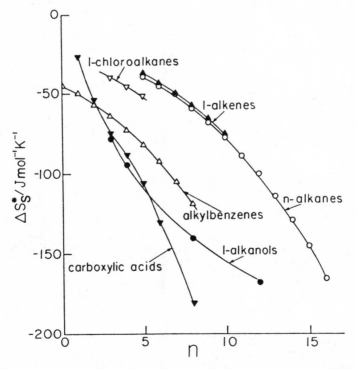

Figure 2.32. Entropy of solvation of hydrocarbon molecules in their own pure liquids as a function of n, the number of carbon atoms in the alkyl chains. All values pertain to $T = 298.15$ K.

as the one presented in Section 2.6). Hence the point corresponding to the solvation Gibbs energy of ethane in hexane will be above the point indicated by the arrow. The larger the solvent n-alkane molecule, the higher the corresponding value of ΔG_s^*, as we indeed observe in Figure 2.31.

In Table 2.20 we summarize some of the thermodynamics of solvation of various hydrocarbons in their own liquids. We note that in order to compute the entropy and enthalpy of solvation, we need to take the derivative of ΔG_s^* with respect to temperature at constant pressure. The available vapor pressure data lie along the liquid–vapor equilibrium line. However, since the vapor pressure is quite low, we can assume that the gaseous phase is ideal. A simple numerical estimate shows that, in this case, the difference between the derivative at constant pressure and along the equilibrium line is negligible. The two derivatives are connected

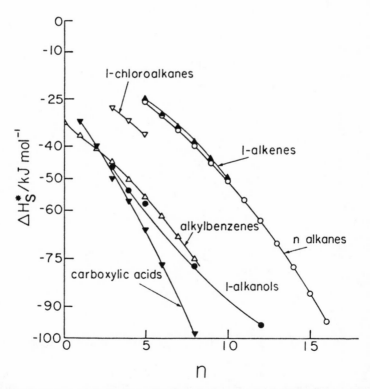

Figure 2.33. Enthalpy of solvation of hydrocarbon molecules in their own pure liquids as a function of n, the number of carbon atoms in the alkyl chains. All values pertain to $T = 298.15$ K.

Table 2.20
Thermodynamic Quantities of Solvation of Organic Molecules in Their Pure Liquids at 298.15 K^a

	$-\Delta G_s^*$ (kJ mol^{-1})	$-\Delta S_s^*$ (JK^{-1} mol^{-1})	$-\Delta H_s^*$ (kJ mol^{-1})
Alkanes			
Cyclopentane	15.928	39.20	27.61
n-Pentane	14.238	39.78	26.14
2-Methylbutane	13.480	37.99	24.81
2,2-Dimethylpropane	11.811	36.19	22.60
Methylcyclopentane	17.568	43.38	30.50
Cyclohexane	18.510	44.39	31.74
n-Hexane	16.952	45.89	30.64
2-Methylpentane	16.100	43.68	29.12
3-Methylpentane	16.409	43.59	29.40
2,2-Dimethylbutane	15.070	40.20	27.05
2,3-Dimethylbutane	15.878	41.79	28.34
n-Heptane	19.639	50.71	35.35
2,4-Dimethylpentane	17.698	47.48	31.86
n-Octane	22.288	60.08	40.20
2,2,4-Trimethylpentane	19.158	49.41	33.89
n-Nonane	25.020	69.11	45.62
2,2,5-Trimethylhexane	21.631	57.99	38.92
n-Decane	27.739	78.61	51.17
n-Undecane	30.539	89.78	57.31
n-Dodecane	33.32	101.01	63.43
n-Tridecane	36.300	115.01	70.58
n-Tetradecane	39.229	129.99	77.98
n-Pentadecane	42.270	146.31	85.88
n-Hexadecane	45.509	165.81	94.94
Alkenes			
1-Pentene	13.819	38.62	25.33
1-Hexene	16.552	44.31	29.75
Cyclohexene	18.911	44.52	32.18
1-Heptene	19.221	50.71	34.34
1-Octene	21.869	57.91	39.13
1-Nonene	24.551	66.19	44.29
1-Decene	27.271	75.90	49.90
Alkylbenzenes			
Benzene	19.058	45.19	32.53
Toluene	21.610	50.29	36.60
Ethylbenzene	23.978	57.11	41.00
1,2-Dimethylbenzene	24.920	58.28	42.30
1,3-Dimethylbenzene	24.309	57.40	41.42
1,4-Dimethylbenzene	24.170	56.90	41.13
n-Propylbenzene	26.229	64.18	45.37
2-Propylbenzene	25.468	62.72	44.16
1,3,5-Trimethylbenzene	27.058	66.11	46.77

(*continued*)

Table 2.20 (*continued*)

	$-\Delta G_s^*$ (kJ mol^{-1})	$-\Delta S_s^*$ (J K^{-1} mol^{-1})	$-\Delta H_s^*$ (kJ mol^{-1})
Alkylbenzenes (*continued*)			
n-Butylbenzene	28.890	72.88	50.63
2,2-Dimethyl-2-propylbenzene	27.019	67.99	47.36
1-(2-propyl)-4-Methybenzene	27.911	69.20	48.54
n-Pentylbenzene	31.472	82.30	56.01
n-Hexylbenzene	34.141	93.01	61.87
n-Heptylbenzene	36.999	105.52	68.45
n-Octylbenzene	39.911	119.70	75.60
Alkanols			
Methanol	20.330	56.69	37.23
Ethanol	21.250	65.48	40.78
1-Propanol	23.258	78.49	46.66
2-Propanol	21.279	73.81	43.28
1-Butanol	25.769	94.89	54.06
2-Butanol	23.070	84.60	49.88
2-Methyl-1-propanol	24.501	96.90	53.39
2-Methyl-2-propanol	20.940	85.40	46.40
1-Pentanol	27.760	102.01	58.17
2-Pentanol	25.489	98.91	54.98
3-Pentanol	24.660	89.58	51.38
3-Methyl-1-butanol	27.719	119.79	63.44
2-Methyl-2-butanol	23.380	107.91	55.55
Cyclohexanol	31.421	100.60	60.01
1-Hexanol	30.011	97.99	59.23
2-Ethyl-1-butanol	28.480	110.92	60.20
1-Heptanol	34.220		
1-Octanol	35.618	141.21	77.72
2-Ethyl-1-hexanol	33.781	138.28	75.01
1-Nonanol	37.911		
1-Decanol	40.082		
1-Dodecanol	45.639	168.11	95.7
Ethers			
Diethyl ether	14.388	42.30	27.00
Dipropyl ether	18.940	46.40	32.77
Di-2-propyl ether	16.840	49.29	31.54
Dibutyl ether	23.899	59.12	41.52
Dipentyl ether	28.438	66.32	48.21
Di-3-methyl-1-butyl ether	27.430	68.28	47.79
1,2-Dimethoxyethane	19.258	36.48	30.14
Bis(2-methoxyethyl)ether	26.170	51.30	41.46
Tetrahydrofuran	17.689	35.90	28.39
Tetrahydropyran	16.309	45.98	30.02
1,4-Dioxane	21.309	42.89	34.10
Ketones			
Acetone	17.560	43.60	30.56
2-Butanone	19.049	45.31	32.56

Table 2.20 (*continued*)

	$-\Delta G_s^*$ (kJ mol^{-1})	$-\Delta S_s^*$ (J K^{-1} mol^{-1})	$-\Delta H_s^*$ (kJ mol^{-1})
Ketones (*continued*)			
2-Pentanone	21.091	55.48	37.64
3-Pentanone	21.012	51.21	36.28
Cyclohexanone	26.179	50.58	41.27
4-Methyl-2-pentanone	20.539	65.60	40.10
2,6-Dimethyl-4-heptanone	26.438	73.39	48.32
Carboxylic acids			
Formic acid	23.171	37.11	34.23
Acetic acid	24.660	54.18	40.82
Propanoic acid	27.719	74.89	50.05
Butanoic acid	30.831	88.62	57.24
Pentanoic acid	34.091	106.40	65.81
Hexanoic acid	37.940	130.92	76.97
Octanoic acid	44.559	181.42	98.65
Esters			
Methyl formate	15.330	28.91	23.94
Ethyl formate	16.869	42.80	29.63
Propyl formate	19.091	46.82	33.05
Butyl formate	21.330	50.50	36.39
2-Methy-1-propyl formate	20.284	46.48	34.12
Methyl acetate	17.309	47.40	31.44
Ethyl acetate	18.840	52.38	34.46
Propyl acetate	20.999	58.28	38.37
Butyl acetate	23.359	64.39	42.56
2-Methyl-1-propyl acetate	22.012	52.01	37.51
Pentyl acetate	25.618	61.80	44.04
Ethyl propanoate	20.761	67.78	38.14
2-Methylpropyl-2-methylpropanoate	24.819	67.78	45.04
Ethyl 3-methylbutanoate	23.890	56.69	40.79
3-Methylbutyl-3-methylbutanoate	28.610	54.39	44.83
Nitriles			
Acetonitrile	20.438	42.22	33.02
Propanonitrile	21.401	45.90	35.09
Butanonitrile	22.610		
Pentanonitrile	24.919	53.60	40.90
Hexanonitrile	26.940	57.32	44.02
Heptanonitrile	27.819		
Octanonitrile	31.400		
Amines			
Propylamine	16.271	47.91	30.55
2-Propylamine	14.639	45.02	28.06
Butylamine	18.798	51.80	34.25
2-Butylamine	17.250	45.48	30.82

(*continued*)

Table 2.20 (*continued*)

	$-\Delta G_s^*$ (kJ mol^{-1})	$-\Delta S_s^*$ (J K^{-1} mol^{-1})	$-\Delta H_s^*$ (kJ mol^{-1})
Amines (*continued*)			
2-Methyl-2-propylamine	15.288	46.32	29.09
Diethylamine	16.430	47.61	30.62
Di-2-propylamine	18.321	46.90	32.30
Dibutylamine	26.710	73.39	48.59
Triethylamine	18.828	49.12	33.47
Cyclohexylamine	24.338	65.90	44.00
Chloroalkanes			
1-Chloropropane	15.890	34.00	27.81
1-Chlorobutane	18.480	45.60	32.07
2-Chlorobutane	17.501	42.38	30.14
1-Chloro-2-methylpropane	17.589	41.30	29.90
2-Chloro-2-methylpropane	15.681	38.12	27.04
1-Chloropentane	21.078	52.09	36.61
Dichloromethane	16.100	40.00	28.02
Chloroform	17.397	40.29	29.53
Carbon tetrachloride	18.409	42.51	31.08
1,1-Dichloroethane	17.02	42.30	29.64
1,2-Dichloroethane	19.648	43.39	32.59
Haloarenes			
Fluorobenzene	19.401	46.48	33.26
2-Fluorotoluene	22.020	51.71	37.43
3-Fluorotoluene	22.250	52.38	37.87
4-Fluorotoluene	22.250	52.30	37.84
Chlorobenzene	23.861	50.79	39.01
1,2-Dichlorobenzene	28.928	62.59	47.53
1,3-Dichlorobenzene	27.949	56.69	44.86
Bromobenzene	26.329	53.89	42.40

[a] Data from Ben-Naim and Marcus.[39]

through the relation

$$\left(\frac{d\Delta G_s^*}{dT}\right)_{eq} = -\Delta S_s^* + \Delta V_s^* \left(\frac{dP}{dT}\right)_{eq} \tag{2.48}$$

where

$$\left(\frac{dP}{dT}\right)_{eq} = \frac{\Delta H_{vap}}{T\Delta V_{vap}} \tag{2.49}$$

The quantity $\Delta H_{vap}/T$ is of the order of 76 J mol^{-1} K^{-1} (Trouton's law) and ΔS_s^* varies between 60 and 160 J mol^{-1} K^{-1}. However, the ratio

$\Delta V_s^*/\Delta V_{vap}$ is very small since ΔV_{vap} is essentially the molar volume of an ideal gas, which is of the order of 2×10^4 cm^3, compared with ΔV_s^*, which is in the range 10^2 to 10^3 cm^3. Hence, the second term on the rhs of equation (2.48) is quite negligible. This conclusion is the same as the one reached in Section 2.4 regarding the inert gas, liquid–vapor equilibrium line.

Figures 2.32 and 2.33 show the dependence of ΔS_s^* and ΔH_s^* on the chain size n of some homologous series. It is discernible that for each

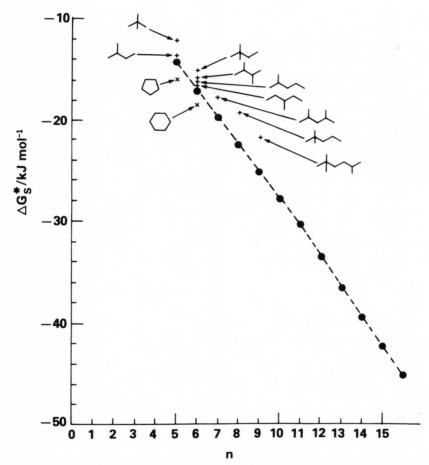

Figure 2.34. Gibbs energy of solvation of alkanes in their own pure liquids as a function of n, the number of carbon atoms in the molecule. Full circles refer to n-alkanes, (+) to branched molecules, and (×) to cyclic molecules. The branched and cyclic molecules are indicated next to the corresponding points.

series the curves of ΔH_s^* and ΔS_s^* as functions of n are very similar. This behavior indicates also that once the Gibbs energy is formed by the combination of $\Delta G_s^* = \Delta H_s^* - T\Delta S_s^*$, the peculiar curvature of each homologous series in the ΔS_s^* and ΔH_s^* curves is removed and the resulting ΔG_s^* curves are almost linear in n.

Finally, it is interesting to note that for all branched alkanes the values of ΔG_s^* are smaller (in absolute magnitude) than the corresponding values for the n-alkane (with the same n). Cyclization, on the other hand, increases the value of ΔG_s^* relative to the n-alkane. This is demonstrated in Figure 2.34.

2.16. SOLVATION OF HYDROCARBONS IN WATER

In this section we report on the solvation thermodynamics of various hydrocarbon molecules in water. Table 2.21 presents some data on ΔG_s^*, based on solubility measurements reported by McAuliffe.[40] Further data on these systems are reported in Table 2.22.[39] Figure 2.35 shows the dependence of ΔG_s^* for some homologous series on the number n of the carbon atoms. Except for the lowest members of the alkane and 1-alkene series, the ΔG_s^* are linear functions of n. It is noteworthy, however, that both the values of ΔG_s^* and their slopes (as functions of n) are positive, in contrast to the negative values of the corresponding quantities for the solvation of hydrocarbons in their own liquids (Section 2.15). Here, in contrast to the discussion given in Section 2.15, we are dealing with different solutes in the *same* solvent. Thus denoting by ΔG_n^* the Gibbs energy of solvation of a molecule with chain length n, we may define

$$\delta G(n + CH_3 \to n + 1) = \Delta G_{n+1}^* - \Delta G_n^* - \Delta G_{CH_3}^* \qquad (2.50)$$

where δG is the indirect work required to bring a methyl group from infinite separation from a hydrocarbon of size n to form a hydrocarbon of size $n + 1$. Since all of the quantities on the rhs of equation (2.50) pertain to the same solvent, the approximate constancy of the slopes of the curves in Figure 2.35 is essentially a reflection of the fact that δG is constant, almost independent of n and of the functional group of a particular homologous series.

Figures 2.36 and 2.37 report the dependence of ΔS_s^* and ΔH_s^* on n. Again, as in Section 2.15, we see that the two sets of curves for ΔH_s^* and ΔS_s^* are similar for each homologous series. It is of interest that the values of ΔS_s^* and ΔH_s^* for ammonia in water fall almost exactly on the curves of the 1-amines series extrapolated to $n = 0$. On the other hand, looking at the series of 1-alkanols we see that the extrapolated values of $n = 0$ do not coincide with the corresponding ΔS_s^* and ΔH_s^* of water in pure water.

Table 2.21
Solvation Gibbs Energies of Hydrocarbons in Water at 298.15 K[a]

Hydrocarbon	Partial pressure (in mm Hg)	ΔG_s^*/kJ mol^{-1}
Methane	736.5	8.083
Ethane	736.5	7.397
Propane	736.5	8.263
n-Butane	736.5	8.987
2-Methylpropane (isobutane)	736.5	9.552
n-Pentane	512.5	9.782
2-Methylbutane (isopentane)	688.0	9.974
2,2-Dimethylpropane (neopentane)	1285.0	11.221
n-Hexane	151.3	10.669
2-Methylpentane (isohexane)	212.0	10.577
3-Methylpentane	190.0	10.493
2,2-Dimethylbutane (neohexane)	320.0	10.886
n-Heptane	45.7	10.991
2,4-Dimethylpentane	98.0	12.070
n-Octane	14.0	12.075
2,2,4-Trimethylpentane (isooctane)	49.0	11.941
2,2,5-Trimethylhexane	16.5	11.393
Cyclopentane	318.0	5.062
Cyclohexane	98.0	5.179
Methylcyclopentane	138.0	6.698
Methylcyclohexane	46.0	7.079
Cycloheptane	22.0	3.355
Cyclo-octane	5.4	3.514
cis-1,2-Dimethylcyclohexane	14.0	6.556
Ethene (ethylene)	736.5	5.305
Propene (propylene)	736.5	5.259
1-Butene	736.5	5.715
2-Methylpropene (isobutylene)	736.5	5.297
1-Pentene	638.0	6.916
2-Pentene	528.0	5.665
3-Methyl-1-butene	736.5	7.594
1-Hexene	186.0	7.004
1-Octene	17.0	9.020
1,3-Butadiene	736.5	2.652
1,4-Pentadiene	736.5	3.907
1,5-Hexadiene	208.0	4.196
1,6-Heptadiene	67.2	5.125
Propyne	736.5	− 2.054
1-Butyne	736.5	− 0.723
1-Pentyne	431.0	0.0167
1-Hexyne	136.0	1.272
1-Heptyne	52.0	2.606
1-Octyne	13.0	2.891
1-Nonyne	6.2	4.338
Benzene	95.2	− 3.694
Toluene	28.5	− 3.200
1,2-Dimethylbenzene (o-xylene)	6.6	− 3.799
Ethylbenzene	9.6	− 2.523
1,2,4-Trimethylbenzene	2.45	− 3.167
2-Propylbenzene	4.6	− 1.284

[a] Based on data from McAuliffe[40] and taken from Ben-Naim and Marcus.[39]

Table 2.22
Solvation Thermodynamical Quantities for Some Solutes in
Water at 298.15 K[a]

Solute	$\Delta G_s^*/$kJ mol^{-1}	$\Delta H_s^*/$kJ mol^{-1}	$T\Delta S_s^*/$kJ mol^{-1}
Helium	11.556	1.54	− 10.02
Neon	11.175	− 1.44	− 12.62
Argon	8.376	− 9.97	− 18.35
Krypton	6.937	− 13.39	− 20.33
Xenon	5.586	− 16.11	− 21.70
Radon	3.678	− 19.05	− 22.73
Hydrogen	9.807	− 1.74	− 11.54
Nitrogen	10.267	− 8.13	− 18.40
Ozone	2.879	− 14.04	− 16.92
Carbon monoxide	9.314	− 8.83	− 18.14
Carbon dioxide	0.469	− 17.48	− 17.95
Methane	8.385	− 11.49	− 19.88
Ethane	7.669	− 17.46	− 25.12
Propane	8.196	− 20.19	− 28.39
n-Butane	8.715	− 23.66	− 32.38
CH_3F	− 0.899	− 15.82	− 14.92
CH_3Cl	− 2.314	− 20.85	− 18.53
CH_3Br	− 3.414	− 23.23	− 19.81
CF_4	+ 13.050	− 12.76	− 25.81
CH_2FCl	− 3.226	− 19.40	− 16.17
CHF_2Cl	+ 0.418	− 26.10	− 26.52
CHF_3	+ 2.757	− 24.42	− 27.17
$CClF_3$	+ 9.334	− 11.10	− 20.44
CCl_2F_2	+ 7.326	− 12.25	− 19.58
H_2S	− 2.276	− 14.99	− 12.71
H_2Se	− 1.774	− 13.40	− 11.62

[a] Based on data from Wilhelm et al.[21]

In Figure 2.38, we present some data on the effect of branching and cyclization on ΔG_s^* of a given alkane. We note that most branched alkanes fall above the curve of the n-alkanes (with the same n), while the cycloalkanes fall below this curve. (A few of the branched alkanes are also below the curve, but in these cases their values are very close to the values of the normal alkanes. In these cases, we judged that the difference between the ΔG_s^* values is quite within the experimental error.)

One can also show that the indirect part of the work required to make a cyclic from a normal alkane is approximately given by

$$\delta G(\text{cyclization}) = \Delta G_{\text{cyclic}}^* - \Delta G_{\text{normal}}^* \qquad (2.51)$$

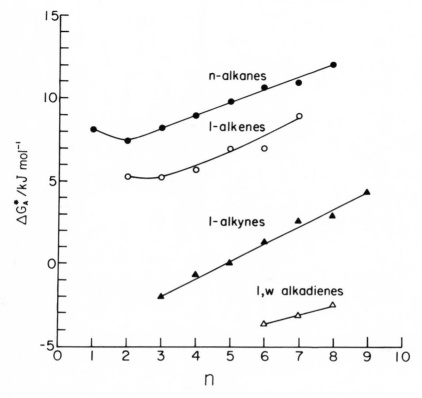

Figure 2.35. Gibbs energy of solvation of some hydrocarbons in water as a function of n, the number of carbon atoms in the alkyl chains. All values pertain to $T = 298.15$ K.

The fact that δG(cyclization) are all negative may be interpreted in terms of hydrophobic interaction between the end methyl groups on the n-alkane (see also Section 3.18).

2.17. SOLVATION OF SIMPLE IONS IN WATER AND AQUEOUS SOLUTIONS

In this section we present some values of solvation thermodynamics of simple ions in water and aqueous solutions. The qualitative basis underlying the method of calculation has been described in Section 1.6, and a statistical mechanical detailed derivation is presented in Section 3.12.

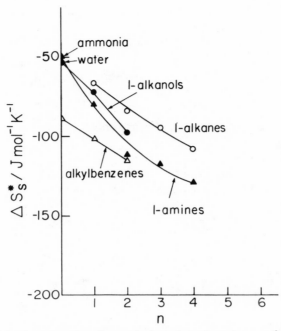

Figure 2.36. Solvation entropies of some hydrocarbons in water as a function of n, the number of carbon atoms in the alkyl chains. All values pertain to $T = 298.15$ K.

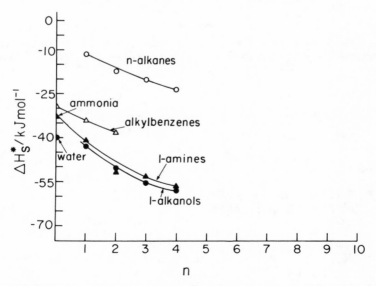

Figure 2.37. Solvation enthalpies of some hydrocarbons in water as a function of n, the number of carbon atoms in the alkyl chains. All values pertain to $T = 298.15$ K.

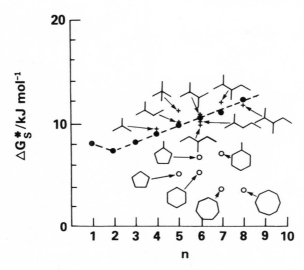

Figure 2.38. Solvation Gibbs energies of alkanes in water as a function of n, the number of carbon atoms in the alkyl chains. All values pertain to $T = 298.15$ K. Branched alkanes are represented by plus (+), cyclic by open circles ($\circ$), and normal alkanes by full circles ($\bullet$).

The general relation between the solvation Gibbs energy of a pair of ions A and B and experimental quantities is

$$\Delta G_{AB}^* = - kT \ln\langle \exp[- \beta B_D(\infty)]\rangle_*$$
$$= - D_0 + kT \ln(\rho_D^g/\rho_A^l\rho_B^l)_{eq} + kT \ln(\Lambda_D^3/\Lambda_A^3\Lambda_B^3 q_{vib}q_{rot}) \quad (2.52)$$

where the meaning of the symbols is discussed in Section 3.12. It is seen that here we write the Gibbs energy of solvation, while in Section 3.12 we have derived an expression for the Helmholtz energy of solvation at T, V constant. These two quantities are equal, as shown in Section 3.1 and 3.12. [Note, however, that for the *same* process, say the solvation at T, P constant, as defined in Section 1.2, the *values* of ΔG_s^* and ΔA_s^* are, in general, different, although this difference might be quite small (Section 3.21).]

The symbol $\langle \quad \rangle_*$ denotes the average over the configurations of *all* the molecules in the system, except the pair of ions which we choose to be our solvaton. In the limit of very dilute solutions, we shall use the symbols ΔG_{AB}^{*0} and $\langle \quad \rangle_0$ to denote the corresponding values in *pure* water.

From equation (2.52) we see that in order to evaluate ΔG_s^* we need some molecular quantities, as well as the densities of the various compo-

Table 2.23

Molecular Properties Used for the Computations in
Tables 2.24 and 2.25[a]

	$\Lambda^3/\text{dm}^3\,\text{mol}^{-1}$	q_{rot}	$D_0'/\text{kJ}\,\text{mol}^{-1}$	ΔE (ionization) $/\text{kJ}\,\text{mol}^{-1}$	$D_0/\text{kJ}\,\text{mol}^{-1}$
H^+	6.1509×10^{-4}				
Cl^-	2.9488×10^{-6}				
Br^-	8.7143×10^{-7}				
I^-	4.3543×10^{-7}				
HCl	2.8274×10^{-6}	19.88	427.72	963.43	1391.15
HBr	8.5520×10^{-7}	24.85	362.60	987.28	1349.88
HI	4.3029×10^{-7}	32.24	294.67	1016.80	1311.47

[a] D_0' is the dissociation energy into atoms. The dissociation energy into ions, D_0, is obtained by adding D_0' and ΔE(ionization), where ΔE(ionization) is the sum of the ionization potential of the cation and the electron affinity of the anion. Values pertain to $T = 298.15$ K. Data taken from Ben-Naim.[46]

nents in the gaseous and liquid phases, at equilibrium. A complete set of such data is available for HCl, HBr, and HI.

The molecular properties required are presented in Table 2.23.[41,42] Here Λ_i^3 is the momentum partition function of species i, defined by

$$\Lambda_i^3 = \left(\frac{h}{\sqrt{2\pi mkT}}\right)^3 \tag{2.53}$$

The rotational partition function is taken as[42]

$$q_{\text{rot}} = T/\theta_r \tag{2.54}$$

where θ_r is the characteristic rotational temperature. The vibrational partition functions for these molecules, at room temperatures, were estimated to be nearly equal to unity (i.e., these molecules are practically in their vibrational ground state at room temperature). The quantity D_0' in Table 2.23 refers to the dissociation of the molecule into two atoms. To obtain the required dissociation energy into ions, D_0, we must add the ionization energy and the electron affinity of the cation and the anion, respectively. The sum of the latter two quantities is denoted by ΔE(ionization); see Tables 2.23 and 2.25.

The number densities in the gaseous phase, ρ_D^g, were computed from the partial vapor pressure over the solution at various compositions. It

Table 2.24
Gibbs Energy of Solvation of HCl, HBr, and HI in Aqueous Solutions of Various Concentrations at 25 °C[a]

x_D^l	ρ_D^l/mol dm^{-3}	ρ_D^g/mol dm^{-3}	ΔG_{AB}^*/kJ mol^{-1}
HCl			
0.083	4.556	2.895×10^{-6}	-1419.5
0.098	5.367	7.935×10^{-6}	-1417.78
0.112	6.122	1.878×10^{-5}	-1416.3
0.126	6.872	4.600×10^{-5}	-1414.65
0.139	7.562	1.030×10^{-4}	-1413.12
0.153	8.304	2.229×10^{-4}	-1411.67
0.165	8.924	4.600×10^{-4}	-1410.24
0.178	9.616	9.059×10^{-4}	-1408.93
0.190	10.210	1.694×10^{-3}	-1407.67
0.201	10.772	3.024×10^{-3}	-1406.50
0.213	11.36	5.306×10^{-3}	-1405.37
	3.699	9.789×10^{-7}	-1421.12
	4.536	2.851×10^{-6}	-1419.48
	5.341	7.530×10^{-6}	-1417.89
	6.122	1.871×10^{-5}	-1416.31
	6.872	4.539×10^{-5}	-1414.68
	7.589	1.038×10^{-4}	-1413.12
	8.287	2.259×10^{-4}	-1411.63
HBr			
0.105	5.556	9.698×10^{-8}	-1388.09
0.120	6.318	2.754×10^{-7}	-1386.14
0.134	7.022	5.535×10^{-7}	-1384.94
0.147	7.659	2.245×10^{-6}	-1381.90
0.151	7.850	2.951×10^{-6}	-1381.34
	5.190	8.122×10^{-8}	-1388.19
	5.921	1.990×10^{-7}	-1386.63
	6.617	4.787×10^{-7}	-1385.00
	7.304	1.216×10^{-6}	-1383.18
	7.932	3.173×10^{-6}	-1381.21
HI			
0.083	4.248	1.608×10^{-5}	-1336.31
0.153	7.2	1.776×10^{-4}	-1332.98
	4.919	3.066×10^{-8}	-1352.56
	5.549	9.789×10^{-8}	-1350.28
	6.129	3.496×10^{-7}	-1347.62
	6.673	1.587×10^{-6}	-1344.29
	7.185	7.100×10^{-6}	-1340.95

[a] x_D^l is the mole fraction of the acid in the solution; ρ_D^l and ρ_D^g are the molar concentrations of the acid in the liquid and gaseous phases. The computations were based on data from Haase $et\ al.$[44] (above the dashed line) and Bates and Kirschman[43] (below the dashed line). Data taken from Ben-Naim.[46]

was assumed that the gaseous phase is ideal, i.e., $\rho_D^g = P/kT$, and that the molecules in the gas are only in the state of dimers. The densities in the liquid $\rho_D^l = \rho_A^l = \rho_B^l$ were computed from the mole fractions of the solutions as reported in the literature.[43,44]

Clearly, this method of evaluating ΔG_{AB}^* is feasible for volatile solutes only. The results pertaining to ΔG_{AB}^* for HCl, HBr, and HI are given in Table 2.24.

Table 2.25

Values of ΔE(ionization) and the Gibbs Energy, Entropy, and Enthalpy of Solvation for Some Simple Electrolytes at 25 °C[a]

	ΔE/kJ mol^{-1} (ionization)	ΔG_{AB}^{*0}/kJ mol^{-1}	ΔH_{AB}^{*0}/kJ mol^{-1}	ΔS_{AB}^{*0}/kJ K^{-1} mol^{-1}
HF	984.18	− 1543.964	− 1608.864	− 0.2177
HCl	963.43	− 1419.422	− 1465.563	− 0.1547
HBr	987.277	− 1392.717	− 1434.104	− 0.1388
HI	1016.8	− 1357.704	− 1389.814	− 0.1077
LiF	192.357	− 968.664	− 1036.935	− 0.2290
LiCl	171.607	− 844.364	− 893.715	− 0.1655
LiBr	195.454	− 817.561	− 862.176	− 0.1496
LiI	224.977	− 782.524	− 817.875	− 0.1186
NaF	167.982	− 863.17	− 922.14	− 0.1978
NaCl	147.232	− 738.643	− 778.92	− 0.1351
NaBr	171.079	− 711.922	− 747.375	− 0.1188
NaI	200.602	− 676.92	− 703.09	− 0.0878
KF	90.94	− 791.337	− 839.302	− 0.1609
KCl	70.19	− 666.787	− 696.082	− 0.0983
KBr	94.037	− 640.09	− 664.532	− 0.0830
KI	123.56	− 605.087	− 620.252	− 0.0509
RbF	75.168	− 768.745	− 813.923	− 0.1515
RbCl	54.418	− 644.195	− 670.713	− 0.0889
RbBr	78.265	− 617.498	− 639.163	− 0.0727
RbI	107.788	− 582.495	− 594.873	− 0.0415
CsF	47.842	− 745.52	− 788.906	− 0.1455
CsCl	27.092	− 621.15	− 645.686	− 0.0823
CsBr	50.939	− 594.263	− 614.136	− 0.0667
CsI	80.462	− 559.26	− 569.856	− 0.0355
AgF	403.13	− 928.217	− 988.832	− 0.2033
AgCl	382.38	− 803.667	− 845.602	− 0.1406
AgBr	406.227	− 776.97	− 814.062	− 0.1244
AgI	435.75	− 741.967	− 769.772	− 0.0933

[a] Data from Ben-Naim.[46]

Table 2.26
Conventional Partial Molar Volumes at Infinite Dilution and Solvation Volumes of Some Simple Electrolytes in Aqueous Solutions at 25 °C[a]

Electrolyte	$\bar{V}_D^0/\text{cm}^3\,\text{mol}^{-1}$	$\Delta V_{AB}^{*0}/\text{cm}^3\,\text{mol}^{-1}$
HF	-1.16	-3.345
HCl	17.83	15.645
HBr	24.71	22.525
HI	36.22	34.035
LiF	-2.04	-4.225
LiCl	16.95	14.765
LiBr	23.83	21.645
LiI	35.34	33.155
NaF	-2.37	-4.555
NaCl	16.62	14.435
NaBr	23.5	21.315
NaI	35.01	32.825
KF	7.86	5.675
KCl	26.85	24.665
KBr	33.73	31.545
KI	45.24	43.055
RbF	12.91	10.725
RbCl	31.9	29.715
RbBr	38.78	36.595
RbI	50.29	48.105
CsF	20.18	17.995
CsCl	39.17	36.985
CsBr	46.05	43.865
CsI	57.56	55.375
AgF	-1.86	-4.045
AgCl	17.13	14.945
AgBr	24.01	21.825
AgI	35.52	33.335

[a] Data from Ben-Naim.[46]

For other salts having negligible partial vapor pressure, the above method is not feasible. However, we could use available tables of standard thermodynamic quantities of formation[45] to convert into solvation quantities in pure water. The details of the conversion procedure are discussed in Sections 3.14 and 3.15. The results obtained for ΔG_{AB}^*, ΔH_{AB}^*, and ΔS_{AB}^* are presented in Table 2.25.

In Table 2.26 we also present some values of the solvation volumes of ionic solutes. These are easily calculated from the corresponding partial molar volumes at infinite dilution:

$$\Delta V^*_{AB} = \bar{V}^0_D - 2kT\beta_T \tag{2.55}$$

where β_T is the isothermal compressibility of the liquid (water). (See Section 3.12.)

Finally, we report on some values of solvation Gibbs energies at highly concentrated aqueous solutions. We employ here the data available on the activity coefficients of the salts in aqueous solutions.[34] The procedure of computation is similar to that described in Section 1.5, but modified to account for the presence of two ions per one molecule D.

Assuming that the ions are structureless particles (i.e., we neglect the contributions due to internal degrees of freedom of the ions), we write the CP of the salt D, at any finite concentration, as

$$\mu^l_D = W(A, B \mid *) + kT \ln \rho_A \rho_B \Lambda^3_A \Lambda^3_B$$
$$= \mu^{o\rho}_D + kT \ln \rho_A \rho_B + 2kT \ln \gamma^\rho_{AB} \tag{2.56}$$

where the first expression on the rhs of relation (2.56) is the statistical mechanical expression. Here $W(A, B \mid *)$ is the coupling work of A and B to the liquid l (which may contain any quantity of the salt D). The second expression on the rhs is the conventional presentation of the CP in terms of the standard CP $\mu^{o\rho}_D$ and the mean activity coefficient γ^ρ_{AB}, both corresponding to the concentration scale ρ.

At infinite dilution relation (2.56) reduces to

$$\mu^l_D = W(A, B \mid 0) + kT \ln \rho_A \rho_B \Lambda^3_A \Lambda^3_B$$
$$= \mu^{o\rho}_D + kT \ln \rho_A \rho_B \tag{2.57}$$

From equation (2.57) we identify the conventional standard CP:

$$\mu^{o\rho}_D = W(A, B \mid 0) + kT \ln \Lambda^3_A \Lambda^3_B \tag{2.58}$$

This is substituted in equation (2.56) to yield

$$\Delta W = W(A, B \mid *) - W(A, B \mid 0) = 2kT \ln \gamma^\rho_{AB} \tag{2.59}$$

Thus from data on the activity coefficient (on the molarity scale) we can compute the difference in the solvation Gibbs energy of the solvaton AB

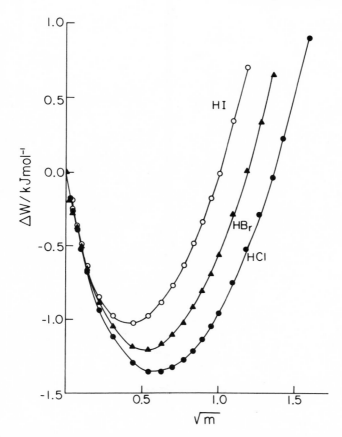

Figure 2.39. Values of $\Delta W = W(A, B\mid *) - W(A, B\mid 0)$ for HCl, HBr, and HI as a function of $\sqrt{m}$, where m is the molality of the acid in the solution. All values pertain to $T = 298.15$ K.

In practice, the mean activity coefficients are given in the molality scale. in a given solution, relative to the solvation Gibbs energy in pure water. The latter may be easily converted to the molarity scale. The procedure of conversion is outlined in Section 3.15. Figures 2.39 to 2.41 present some values of $\Delta W = W(A, B\mid *) - W(A, B\mid 0)$ as functions of the solute concentration in aqueous solutions. Some further details on the computational procedure may be found elsewhere.[46]

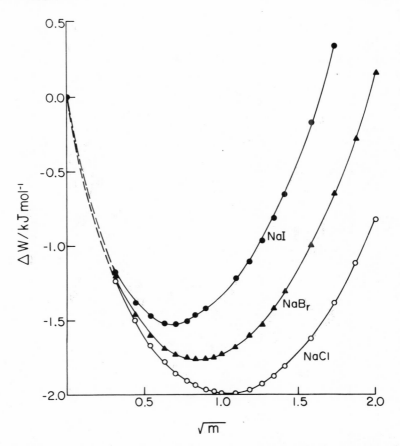

Figure 2.40. Values of $\Delta W = W(A, B\,|\,*) - W(A, B\,|\,0)$ for NaCl, NaBr, and NaI as a function of $\sqrt{m}$, where m is the molality of the salt in the solution. All values pertain to $T = 298.15$ K.

2.18. SOLVATION OF PROTEINS IN AQUEOUS SOLUTIONS

The fundamental relation between solvation Gibbs energy of a molecule s in a liquid phase l is

$$\Delta G_s^{*l} = kT \ln(\rho_s^{ig}/\rho_s^{l})_{\text{eq}} \qquad (2.60)$$

where ρ_s^{ig} and ρ_s^{l} are the number densities of s in an ideal gas and in the liquid phase at equilibrium.

Of course, this relation is impractical for calculating the Gibbs energy of solvation of solutes for which the vapor pressure, or the density in the gaseous phase, is unmeasurable. In such cases, the most we can do is to measure differences in the solvation Gibbs energy of s between two phases [see equation (1.22), Section 1.3], i.e.,

$$\Delta(\Delta G_s^*) = \Delta G_s^{*\beta} - \Delta G_s^{*\alpha} = kT \ln(\rho_s^\alpha/\rho_s^\beta)_{eq} \tag{2.61}$$

where $(\rho_s^\alpha/\rho_s^\beta)_{eq}$ is the equilibrium density ratio of s in the two phases α and β. Relation (2.61) may be applied for aqueous protein solutions.

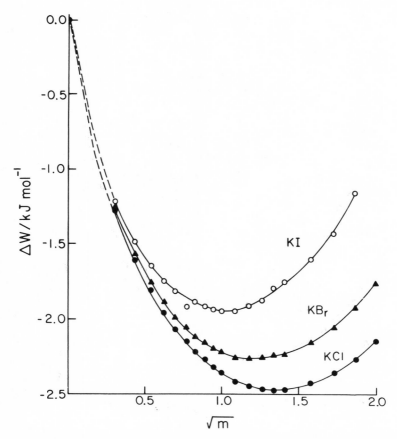

Figure 2.41. Values of $\Delta W = W(A, B\,|\,*) - W(A, B\,|\,0)$ for KCl, KBr, and KI in aqueous solutions as a function of $\sqrt{m}$, where m is the molality of the salt in the solution. All values pertain to $T = 298.15$ K.

We present in this section one series of results due to Shanbhag and Axelsson,[47] who were interested primarily in establishing the relative "hydrophobicity" of various proteins (for further discussion of the term hydrophobicity see Section 3.19).

In this study the density ratio of a protein was measured in two phases. These two phases are essentially aqueous soutions, and they differ in what may be referred to as the "hydrophobicity content" of the solution. The two phases are described in Figure 2.42. In the first system, I, the two solutions contain polyethylene glycol (PEG) and dextran in different weight percentages as indicated in the figure. In a second system, II, part of the PEG is replaced by a fatty acid ester of polyethylene glycol (P-PEG) in such a way that the total weight percentage of the dextran and the PEG + P-PEG is the same as in system I.

The structural formula of the fatty acid ester of PEG is

$$\underbrace{CH_3-(CH_2)_x-\overset{\displaystyle O}{\overset{\displaystyle \|}{C}}-O}_{\text{Fatty acid}}\underbrace{-CH_2-CH_2(-O-CH_2-CH_2-)_n-OH}_{\text{PEG}} \qquad (2.62)$$

with $x = 0, 2, 4, \ldots, 16$ and $\bar{n} = 135$.

Clearly, if we vary x, i.e., the length of the fatty acid residue, we make the P-PEG more hydrophobic (note that P-PEG refers specifically to palmitate-PEG, but we shall use this abbreviation to indicate any fatty acid ester of PEG).

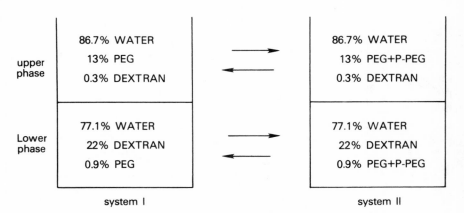

Figure 2.42. Schematic illustration of the two systems used for the measurement of the hydrophobicity of various proteins.

If we measure the density ratio for a given protein between the lower (l) and upper (u) solutions, we have, for each of the systems shown in Figure 2.42, the difference in the corresponding solvation Gibbs energies:

$$\Delta(\Delta G_s^{*I})(u \to l) = \Delta G_s^{*l,I} - \Delta G_s^{*u,I} = kT \ln(\rho_s^{u,I}/\rho_s^{l,I})_{eq} \qquad (2.63)$$

$$\Delta(\Delta G_s^{*II})(u \to l) = \Delta G_s^{*l,II} - \Delta G_s^{*u,II} = kT \ln(\rho_s^{u,II}/\rho_s^{l,II})_{eq} \qquad (2.64)$$

These two are measurable quantities. Taking the difference between the two, we obtain

$$\Delta(\Delta G_s^*) = \Delta(\Delta G_s^{*I}) - \Delta(\Delta G_s^{*II}) = (\Delta G_s^{*l,I} - \Delta G_s^{*l,II}) + (\Delta G_s^{*u,II} - \Delta G_s^{*u,I})$$

$$= \Delta(\Delta G_s^{*l})(II \to I) + \Delta(\Delta G_s^{*u})(I \to II) \qquad (2.65)$$

where we have rewritten the quantity $\Delta(\Delta G^*)$ (difference of difference of solvation Gibbs energies) as a sum of the solvation Gibbs energy of transfer of s from II to I between the lower solutions, and the solvation Gibbs energy of transfer of s from I to II between the upper solutions.

The authors used this quantity as their operational definition of the hydrophobicity of the protein s.

Strictly speaking, the quantity defined in relation (2.65) is a difference of differences in solvation Gibbs energy. However, if we examine the content of the solutions in the diagram (Figure 2.42), we see that the concentration of PEG and P-PEG in the lower solutions is quite small relative to the concentrations in the upper solutions. Therefore, as a first approximation we can assume that $\Delta(\Delta G_s^{*l})(II \to I)$ in relation (2.65) is relatively small compared to $\Delta(\Delta G_s^{*u})(II \to I)$. Hence we write the approximation

$$\Delta(\Delta G_s^*) \approx \Delta\Delta G_s^{*u}(I \to II) = \Delta G_s^{*u}(II) - \Delta G_s^{*u}(I) \qquad (2.66)$$

We note that the only difference between the two upper phases is the addition of the fatty acid residue, which makes solution II more hydrophobic relative to I. When we transfer a protein which has more hydrophobic regions (see also Section 3.19) from I to II, we expect that $\Delta(\Delta G^*)$ will be more negative. Thus, for a fixed fatty acid, different proteins will have different affinities to P-PEG, relative to PEG. Figure 2.43 shows some results for four proteins and for a series of chain lengths of the fatty acids. It is observed that for short chain lengths ($n \lesssim 8$) there is almost no noticeable effect on $\Delta(\Delta G_s^*)$. This behavior is probably due to the weak exposure of the fatty acids from the PEG polymer. However, for $n \gtrsim 8$ we observe a dramatic dependence of $\Delta(\Delta G_s^*)$ on the chain length n. Note that for cytochrome c we observe almost no variation of $\Delta(\Delta G_s^*)$

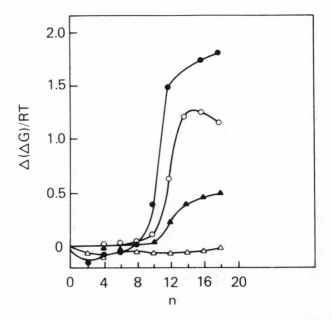

Figure 2.43. Effect of the chain length of the polyethylene-glycol-bound aliphatic ligand, n, on the change in the partition coefficient of the protein as expressed by $\Delta(\Delta G)$ defined in equation (2.66). The proteins are human serum albumin (●), β-lactoglobulin (○), CO-hemoglobin (▲), and cytochrome c (△). Redrawn from Shanbhag and Axelsson.[47]

as a function of n. This is consistent with the well-known fact[48] that in cytochrome c most of the amino acids on its surface are hydrophilic. In contrast, human serum albumin is known to have a relatively high content of hydrophobic amino acids on its surface (which makes it an effective carrier of fatty acids in the blood). We observe a very sharp increase in $-\Delta(\Delta G_s^*)$ as a function of n. Thus, if we take a fixed value of n, say $n = 15$, we may construct a hydrophobicity scale for the various proteins according to their corresponding values of $\Delta(\Delta G_s^*)$.

The above set of results demonstrates the applicability of the solvation concept to protein solutions. Although we might not be able to measure ΔG_s^{*l} for a protein in a liquid, it should be feasible to find a standard solvent α and use equation (2.61) to compare various proteins relative to this standard solvent.

Chapter 3

Further Theoretical Background

3.1. THE STATISTICAL MECHANICAL EXPRESSION OF THE CHEMICAL POTENTIAL IN VARIOUS ENSEMBLES

In Section 1.1 we presented the fundamental expression for the chemical potential (CP) and discussed some of its features. This expression constitutes the cornerstone of the entire book. In this section we present a detailed derivation of this expression using three different ensembles or sets of independent variables: T, V, N; T, P, N; and T, V, μ.

(1) *The T, V, N ensemble.* For simplicity we derive here the required expression for the CP for a one-component system of simple spherical particles. We shall point out some possible generalizations at the end of this section.

Consider a system characterized by the variables T, V, N, i.e., N particles contained in a volume V and maintained at a temperature T. The CP in such a system is defined as

$$\mu = \left(\frac{\partial A}{\partial N}\right)_{T,V} = \lim_{\Delta N \to 0} \frac{A(T, V, N + \Delta N) - A(T, V, N)}{\Delta N}$$

$$= \lim_{\Delta N \to 0} \left[A\left(T, \frac{V}{\Delta N}, \frac{N}{\Delta N} + 1\right) - A\left(T, \frac{V}{\Delta N}, \frac{N}{\Delta N}\right) \right]$$

$$= \lim_{\substack{V' \to \infty \\ N' \to \infty}} [A(T, V', N' + 1) - A(T, V', N')] \tag{3.1}$$

The first two equalities on the rhs of relation (3.1) follow from the definition of the derivative with respect to the variable N.

The third equality follows from the extensive character of the Helmholtz energy. The last equality states that instead of fixing V and N and adding the infinitesimal quantity ΔN, we may let V' and N' increase to infinity, keeping their ratio N'/V' fixed, and adding unity to N' to effect an infinitesimal change in the variable N. We shall use the last equality in what follows.

The fundamental link between the thermodynamic quantity A and statistical mechanics is through the relation

$$A = -kT \ln Q \tag{3.2}$$

where $Q(T, V, N)$ is the so-called canonical partition function. For our simple system this partition function has the form[49]

$$Q(T, V, N) = \frac{q^N}{N! \Lambda^{3N}} \int \cdots_V \int d\mathbf{R}_1 \cdots d\mathbf{R}_N \exp[-\beta U_N(\mathbf{R}_1 \cdots \mathbf{R}_N)] \tag{3.3}$$

Here q is the internal partition function of a single particle. It may include contributions from rotational, vibrational, electronic, and nuclear degrees of freedom. In equation (3.3) we have assumed that all these internal degrees of freedom are separable, i.e., they are independent of the environment surrounding each molecule. If these do depend on the configuration of the system then they should be included under the integral sign.[†]

The quantity Λ^3 is the momentum partition function. It arises from the integration over all possible translational momenta of the molecule. It depends on the molecular mass m, and the temperature T, as follows:

$$\Lambda^3 = \frac{h^3}{(2\pi m k T)^{3/2}} \tag{3.4}$$

where h is the Planck constant and k the Boltzmann constant; Λ has the dimension of length and is sometimes also referred to as the De Broglie thermal wavelength, $\mathbf{R}_i$ is the locational vector describing the location of the center of the ith molecule, while $d\mathbf{R}_i$ stands for an element of volume $dx_i dy_i dz_i$ located at the center of the ith molecule. The integration in equation (3.3) is extended for each molecule, over the entire volume V of the system. The symbol $U_N(\mathbf{R}_1 \cdots \mathbf{R}_N)$ denotes the total potential energy

---

† In more general cases one should start with the fully quantum-mechanical partition function. Here we use only its classical limit.

of the N molecules at the specific configuration $\mathbf{R}_1 \cdots \mathbf{R}_N$. The factor $N!$ arises from the indistinguishability of the N molecules, and $\beta = (kT)^{-1}$.

We now construct the expression for the CP by taking the ratio of the canonical partition functions for $N + 1$ and N particles. Thus, using equations (3.1), (3.2), and (3.3), we have

$$
\begin{aligned}
\mu &= A(N + 1) - A(N) \\
&= -kT \ln[Q(N + 1)/Q(N)] \\
&= -kT \ln \left\{ \left[\frac{q^{N+1}}{(N + 1)! \, \varLambda^{3(N+1)}} \int \cdots \int d\mathbf{R}_0 d\mathbf{R}_1 \cdots d\mathbf{R}_N \right. \right. \\
&\qquad\qquad \left. \times \{\exp[-\beta U_{N+1}(\mathbf{R}_0 \cdots \mathbf{R}_N)] \} \right] \\
&\qquad \left. \times \left\{ \frac{q^N}{N! \, \varLambda^{3N}} \int \cdots \int d\mathbf{R}_1 \cdots d\mathbf{R}_N \exp[-\beta U_N(\mathbf{R}_1 \cdots \mathbf{R}_N)] \right\}^{-1} \right\} \\
&= -kT \ln \left\{ \frac{q}{(N + 1)\varLambda^3} \int \cdots \int d\mathbf{R}_0 \cdots d\mathbf{R}_N \right. \\
&\qquad\qquad \left. \times P(\mathbf{R}_1 \cdots \mathbf{R}_N) \exp[-\beta B_0(\mathbf{R}_0 \cdots \mathbf{R}_N)] \right\}
\end{aligned}
\tag{3.5}
$$

In the last form on the rhs of equation (3.5) we used the notation $P(\mathbf{R}_1 \cdots \mathbf{R}_N)$ for the fundamental probability density of finding the configuration $\mathbf{R}_1 \cdots \mathbf{R}_N$ of the N particles:

$$
P(\mathbf{R}_1 \cdots \mathbf{R}_N) = \frac{\exp[-\beta U_N(\mathbf{R}_1 \cdots \mathbf{R}_N)]}{\displaystyle\int \cdots \int d\mathbf{R}_1 \cdots d\mathbf{R}_N \exp[-\beta U_N(\mathbf{R}_1 \cdots \mathbf{R}_N)]}
\tag{3.6}
$$

Also, the total interaction energy among the $N + 1$ molecules has been written as

$$
U_{N+1}(\mathbf{R}_0 \mathbf{R}_1 \cdots \mathbf{R}_N) = U_N(\mathbf{R}_1 \cdots \mathbf{R}_N) + B_0(\mathbf{R}_0 \cdots \mathbf{R}_N)
\tag{3.7}
$$

Equation (3.7) may be viewed as a definition of the *binding energy*, B_0, of the newly added particle at $\mathbf{R}_0$ to all the N other particles of the system being at a specific configuration $\mathbf{R}_1 \cdots \mathbf{R}_N$.

Because of the invariance of the integral in equation (3.5) to the choice of the origin of our coordinate system, we can measure all the vectors $\mathbf{R}_i$ with respect to an origin chosen at $\mathbf{R}_0$. By effecting this change of variables (details are given elsewhere[4]), we may transform equation

(3.5) into a simpler form, i.e.,

$$\mu = -kT \ln \left\{ \frac{qV}{(N+1)\Lambda^3} \int \cdots \int d\mathbf{R}_1' \cdots d\mathbf{R}_N' P(\mathbf{R}_1' \cdots \mathbf{R}_N') \right.$$

$$\left. \times \exp[-\beta B_0(\mathbf{R}_1' \cdots \mathbf{R}_N')] \right\} \tag{3.8}$$

where the two functions in the integrand depend on the (relative) coordinates $\mathbf{R}_1' \cdots \mathbf{R}_N'$. The integration over $\mathbf{R}_0$ produced the volume V. We may now put $\rho = (N+1)/V \cong N/V$ (macroscopic system) and rewrite equation (3.8) in the more compact form

$$\mu = -kT \ln \left[\frac{q}{\rho \Lambda^3} \langle \exp(-\beta B_0) \rangle \right] \tag{3.9}$$

where the symbol $\langle \quad \rangle$ signifies an average over all the configurations of the N particles (excluding the newly added one) using the probability density $P(\mathbf{R}_1 \cdots \mathbf{R}_N)$ of equation (3.6).

We now repeat the same derivation as above, but with one additional restriction: that the newly added particle be placed at a fixed position $\mathbf{R}_0$. The pseudo-CP (PCP), as defined in Section 1.1, is thus obtained in the form

$$\mu^* = A(N+1, \mathbf{R}_0) - A(N)$$

$$= -kT \ln \left\{ \frac{\frac{q^{N+1}}{N! \Lambda^{3N}} \int d\mathbf{R}_1 \cdots d\mathbf{R}_N \exp[-\beta U_{N+1}(\mathbf{R}_0 \cdots \mathbf{R}_N)]}{\frac{q^N}{N! \Lambda^{3N}} \int d\mathbf{R}_1 \cdots d\mathbf{R}_N \exp[-\beta U_N(\mathbf{R}_1 \cdots \mathbf{R}_N)]} \right\}$$

$$= -kT \ln \left\{ q \int \cdots \int d\mathbf{R}_1 \cdots d\mathbf{R}_N P(\mathbf{R}_1 \cdots \mathbf{R}_N) \exp[-\beta B_0(\mathbf{R}_1 \cdots \mathbf{R}_N)] \right\}$$

$$= -kT \ln[q \langle \exp(-\beta B_0) \rangle] \tag{3.10}$$

It is important to examine the differences between equation (3.5) and equation (3.10). Understanding the origin of these differences is crucial for understanding the approach to solvation thermodynamics as advocated in this book.

First, in equation (3.5) we have integration over $N+1$ locations versus N locations in equation (3.10). This produces the factor V in equation (3.8) which is absent from equation (3.10). In relation (3.5) we have $(N+1)$ momentum partition functions Λ^3 versus N in relation

(3.10). Finally, in equation (3.5) we have $(N + 1)!$ compared with $N!$ in equation (3.10). All these differences arise from the fact that in equation (3.10) we have constrained the $(N + 1)$th particle to a fixed position. Such a particle cannot wander in the entire volume V, does not have momentum partition function Λ^3, and, being at a fixed position, is *distinguishable* from the rest of the N members of the same species.

Combining equation (3.9) with equation (3.10), we arrive at our fundamental expression

$$\mu = \mu^* + kT \ln \rho \Lambda^3 \qquad (3.11)$$

which corresponds to a splitting of the process of adding one particle to the system (at T, V constant) in two steps. First, we place the particle at a fixed position, say $\mathbf{R}_0$. The corresponding change in the Helmholtz energy is μ^*. Next, we release the constraint imposed on the particle; this step always lowers the Helmholtz energy (since our system is classical $\rho \Lambda^3 \ll 1$) by the amount $kT \ln \rho \Lambda^3$, to which we referred as the *liberation Helmholtz energy*.

We can now also appreciate the various ingredients that contribute to the last term. The particle that is released from its fixed position may now wander in the entire volume V. This factor arises from the $N + 1$ integrals in equation (3.5). Second, the released particle gains momentum and hence acquires a momentum partition function Λ^3. Finally, as long as the new particle is at a fixed position, it is distinguishable from all the other particles. Once it has been released, it becomes indistinguishable from the rest of the N particles. We say that this particle is being *assimilated* by the other members of the same species.[†] Altogether, these three factors form the liberation Helmholtz energy.

(2) *The T, P, N ensemble*. We now derive the expression for the CP in the T, P, N ensemble. The set of variables T, P, N is the most important one in actual experiments. The derivation here will be quite condensed, since most of the steps are similar to those carried out in the T, V, N ensemble. A more detailed derivation may be found elsewhere.[(4)]

For a one-component system the CP in the T, P, N ensemble is defined by

$$\mu = (\partial G/\partial N)_{T,P} = G(T, P, N + 1) - G(T, P, N) \qquad (3.12)$$

As in the previous derivation, we first express the CP in terms of the

---

[†] The word *assimilation* is used here to distinguish it from the process of *mixing*. The latter is used whenever two distinguishable species are mixed. For more details see Chapter 4.

ratio of the corresponding partition functions. This gives the result

$$
\mu = -kT \ln \left\{ \left[q \int dV \int d\mathbf{R}_0 \cdots d\mathbf{R}_N \exp[-\beta U_{N+1}(\mathbf{R}_0 \cdots \mathbf{R}_N) - \beta PV] \right] \right.
$$

$$
\times \left[\Lambda^3(N+1) \int dV \int d\mathbf{R}_1 \cdots d\mathbf{R}_N \right.
$$

$$
\left. \left. \times \exp[-\beta U_N(\mathbf{R}_1 \cdots \mathbf{R}_N) - \beta PV] \right]^{-1} \right\} \tag{3.13}
$$

We now use the conditional distribution function[†] of finding a configuration $\mathbf{R}_1 \cdots \mathbf{R}_N$, given that the system has a volume V, i.e.,

$$
P(\mathbf{R}_1 \cdots \mathbf{R}_N/V) = \frac{\exp[-\beta U_N(\mathbf{R}_1 \cdots \mathbf{R}_N)]}{\int \cdots_V \int d\mathbf{R}_1 \cdots d\mathbf{R}_N \exp[-\beta U_N(\mathbf{R}_1 \cdots \mathbf{R}_N)]} \tag{3.14}
$$

Using this distribution function, equation (3.13) may be rewritten as

$$
\mu = -kT \ln \left[\frac{q}{\Lambda^3(N+1)} \int_0^\infty dV\, P(V) \int_V d\mathbf{R}_0 \cdots d\mathbf{R}_N \right.
$$

$$
\left. \times P(\mathbf{R}_1 \cdots \mathbf{R}_N/V) \exp(-\beta B_0) \right] \tag{3.15}
$$

The essential difference between equations (3.15) and (3.5) is the additional integration over all possible volumes, with the probability distribution $P(V)$. Assuming that the system is macroscopically large, the probability density $P(V)$ should have a single sharp peak at $\langle V \rangle$. For simplicity we may assume that $P(V)$ behaves as a Dirac delta function:

$$
P(V) = \delta(V - \langle V \rangle) \tag{3.16}
$$

where $\langle V \rangle$ is the average volume of the system characterized by the variables T, P, N.

This assumption enables us to simplify expression (3.15) as follows:

$$
\mu = -kT \ln \left[\frac{q\langle V \rangle}{\Lambda^3(N+1)} \int \cdots_{\langle V \rangle} \int d\mathbf{R}_1 \cdots d\mathbf{R}_N \right.
$$

$$
\left. \times P(\mathbf{R}_1 \cdots \mathbf{R}_N/\langle V \rangle) \exp(-\beta B_0) \right]
$$

$$
= -kT \ln \left[\frac{q}{\rho \Lambda^3} \langle \exp(-\beta B_0) \rangle \right] \tag{3.17}
$$

[†] Note that P denotes the pressure in equation (3.13) and a distribution function in relation (3.14).

which is identical in form with expression (3.9). The difference is in the definition of the density $\rho = (N + 1)/\langle V \rangle$, and the meaning of the average sign $\langle \quad \rangle$ which, in relation (3.17), is over all configurations of the N particles with the conditional distribution given in expression (3.14).

We can repeat exactly the same procedure, but instead of adding a free particle to the system we place it at a fixed position (within the boundaries of the system). In this way we derive the corresponding expression for the PCP:

$$\mu^* = -kT \ln[q \langle \exp(-\beta B_0) \rangle] \tag{3.18}$$

The last two equations are now combined to yield

$$\mu = \mu^* + kT \ln \rho \Lambda^3 \tag{3.19}$$

which is identical in form with equation (3.11), the corresponding equation in the T, V, N ensemble.

Now, from the equalities

$$\mu(T, P, N) = \left(\frac{\partial G}{\partial N}\right)_{P,T} = G(T, P, N + 1) - G(T, P, N) \tag{3.20}$$

and

$$\mu(T, V, N) = \left(\frac{\partial A}{\partial N}\right)_{V,T} = A(T, V, N + 1) - A(T, V, N) \tag{3.21}$$

it follows that if we choose the systems characterized by the variables T, P, N and T, V, N, respectively, such that the average volume $\langle V \rangle$ in the former is equal to the volume of the latter, the two CPs in equations (3.20) and (3.21) will have equal values. (The same is true if we require that the pressure in the T, V, N ensemble be the same as the pressure in the T, P, N ensemble.)

If we require that $V = \langle V \rangle$, then also the densities $\rho = N/V$ in equation (3.11) and $\rho = N/\langle V \rangle$ in equation (3.19) wil be the same. Hence, the liberation Helmholtz energy in equation (3.11) and the liberation Gibbs energy in equation (3.19) are equal, and therefore also the PCPs, namely

$$\mu^*(T, V, N) = \mu^*(T, P, N) \tag{3.22}$$

with the requirement that $V = \langle V \rangle$. Since the PCP in an ideal-gas phase is zero (for the case of particles without internal degrees of freedom), it also follows that the Gibbs solvation energy (at T, P, N constant) is equal to the Helmholtz solvation energy (at T, V, N with $V = \langle V \rangle$), i.e.,

$$\Delta G^*(T, P, N) = \Delta A^*(T, V, N) \qquad \text{with } V = \langle V \rangle \qquad (3.23)$$

We note, however, that ΔG^* and ΔA^* in this equation *do not* pertain to the *same* process of sovation. If we restrict ourselves to the *same* process, say at T, P, N constant, then we have the equality

$$\Delta G^*(T, P, N) = \Delta A^*(T, P, N) + P\Delta V^*(T, P, N) \qquad (3.24)$$

Thus ΔG^* and ΔA^* for the *same* process are, in general, different quantities. In all our examples the difference is quite small, as shown in Section 3.21.

(3) *The T, V, μ ensemble.* For completeness we add here one more expression for the CP in an open system characterized by the variables T, V, μ. Here, the CP is one of the independent variables used to specify our system. The corresponding partition function — the so-called Grand partition function — is given for a one-component classical system by[49] (see also Chapter 3 in Ben-Naim[4])

$$\Xi(T, V, \mu) = \sum_{N \geq 0} \frac{z^N}{N!} \int \cdots \int d\mathbf{R}^N \exp[-\beta U_N(\mathbf{R}^N)] \qquad (3.25)$$

where $z = \exp(\beta\mu)/\Lambda^3$, and $\mathbf{R}^N$ is a shorthand notation for $\mathbf{R}_1 \cdots \mathbf{R}_N$.

Likewise, we write a Grand partition function characterized by the same variables T, V, μ, but with the additional particle at a fixed position, say $\mathbf{R}_0$ within the volume V, i.e,

$$\Xi(T, V, \mu; \mathbf{R}_0) = \sum_{N \geq 0} \frac{z^N}{N!} \int \cdots \int d\mathbf{R}^N \exp[-\beta U_{N+1}(\mathbf{R}^N, \mathbf{R}_0)] \qquad (3.26)$$

We note that equation (3.26) may be viewed as a partition function for a system subjected to an "external" field of force created by a particle placed at $\mathbf{R}_0$.

The average number of particles in the open system is given by the well-known relation[49]

$$\langle N \rangle = kT \left(\frac{\partial \ln \Xi}{\partial \mu}\right)_{T,V}$$
$$= \frac{z}{\Xi} \sum_{N \geq 1} \frac{Nz^{N-1}}{N!} \int \cdots \int d\mathbf{R}^N \exp[-\beta U_N(\mathbf{R}^N)] \qquad (3.27)$$

Now, each integral on the rhs of equation (3.27) may be rewritten as

$$\int \cdots \int d\mathbf{R}^N \exp[-\beta U_N(\mathbf{R}^N)] = V \int d\mathbf{R}^{N-1} \exp[-\beta U_N(\mathbf{R}^{N-1}, \mathbf{R}_1)] \tag{3.28}$$

where we have transformed to coordinates relative to, say, $\mathbf{R}_1$ and integrated over $\mathbf{R}_1$ to obtain the volume. From equations (3.27) and (3.28) we obtain

$$
\begin{aligned}
\langle N \rangle &= \frac{zV}{\Xi} \sum_{N \geq 1} \frac{z^{N-1}}{(N-1)!} \int \cdots \int d\mathbf{R}^{N-1} \exp[-\beta U_N(\mathbf{R}^{N-1}, \mathbf{R}_1)] \\
&= \frac{zV}{\Xi} \sum_{N \geq 0} \frac{z^N}{N!} \int \cdots \int d\mathbf{R}^N \exp[-\beta U_{N+1}(\mathbf{R}^N, \mathbf{R}_1)] \\
&= \frac{zV\Xi(T, V, \mu; \mathbf{R}_1)}{\Xi(T, V, \mu)}
\end{aligned}
\tag{3.29}
$$

where we have changed the summation index and identified the resulting sum with equation (3.26). [Note that the choice of the fixed position $\mathbf{R}_0$ in equation (3.26) or $\mathbf{R}_1$ in equation (3.29) produces the same quantities $\Xi(T, V, \mu, \mathbf{R}_0) = \Xi(T, V, \mu, \mathbf{R}_1)$.]

Equation (3.29) may now be rewritten as

$$
\begin{aligned}
\mu &= kT \ln \rho \Lambda^3 - [kT \ln \Xi(T, V, \mu; \mathbf{R}_0) - kT \ln \Xi(T, V, \mu)] \\
&= kT \ln \rho \Lambda^3 - [P(T, V, \mu; \mathbf{R}_0)V - P(T, V, \mu)V]
\end{aligned}
\tag{3.30}
$$

where the second term on the rhs is the (PV) work required to place a particle at a fixed position $\mathbf{R}_0$ in the system. The density ρ is defined as $\rho = \langle N \rangle / V$. Thus relation (3.30) has the same form as our basic expression (3.11), except for a reinterpretation of the density and the work term according to the specific variables used to characterize our system.

The second term on the rhs of relation (3.30) may be rewritten, noting 3.25 and 3.26, in the form

$$\mu = kT \ln \rho \Lambda^3 - kT \ln \langle \exp(-\beta B_0) \rangle \tag{3.31}$$

which has the same form as equation (3.9) with the appropriate reinterpretation of the average quantities in the T, V, μ ensemble.

Finally, we note that by a straightforward generalization of the argument given above one may obtain essentially the same expressions for the CP of a nonspherical particle and in multicomponent systems.

Therefore, the only restriction that we still require is that our systems obey the classical limit of statistical mechanics.

3.2. THE SOLVATION HELMHOLTZ ENERGY OF A MOLECULE HAVING INTERNAL ROTATIONAL DEGREES OF FREEDOM

Let s be a molecule with internal rotational degrees of freedom. We assume that the vibrational, electronic, and nuclear partition functions are separable and independent of the configuration of the molecules in the system. We define the PCP of a molecule with a fixed conformation $\mathbf{P_s}$ as the change in the Helmholtz energy for the process of introducing s into the system l (at fixed T, V), in such a way that its center of mass is at a fixed position $\mathbf{R_s}$, its conformation $\mathbf{P_s}$ is "frozen in," and the orientation of the entire molecule $\mathbf{\Omega_s}$ is also fixed. If we now release the constraint on the fixed position of the center of mass, we may define the chemical potential of the $\mathbf{P_s}$ conformer in the gas and liquid phases as follows:

$$\mu_s^g(\mathbf{P_s}) = \mu_s^{*g}(\mathbf{P_s}) + kT \ln \rho_s^g \Lambda_s^3 \qquad (3.32)$$

$$\mu_s^l(\mathbf{P_s}) = \mu_s^{*l}(\mathbf{P_s}) + kT \ln \rho_s^l \Lambda_s^3 \qquad (3.33)$$

We note that the rotational partition function of the entire molecule as well as the internal partition functions of s are included in the PCP. In classical systems the momentum partition function Λ_s^3 is independent of the environment, whether it is a gas or a liquid phase.

The solvation Helmholtz energy of the $\mathbf{P_s}$ conformer is defined as

$$\Delta\mu_s^*(\mathbf{P_s}) = \mu_s^{*l}(\mathbf{P_s}) - \mu_s^{*g}(\mathbf{P_s}) = -kT \ln \langle \exp[-\beta B_s(\mathbf{P_s})] \rangle \qquad (3.24)$$

i.e., this is the Helmholtz energy of transferring an s molecule, being frozen in its $\mathbf{P_s}$ conformation, from a fixed position in g into a fixed position in l. Clearly, since $\mathbf{P_s}$ is fixed, the orientation of the entire molecule $\mathbf{\Omega_s}$ does not affect the solvation Holmholtz energy. If we also assume that all vibrational, electronic, and nuclear degrees of freedom are not affected by this transfer from g to l, we can write the second equality in equation (3.34), where we have the same average quantity as in equation (3.9), with the additional constraint that the conformation of the molecule, $\mathbf{P_s}$, is frozen in.

Next, we wish to find the relation between $\Delta\mu_s^*(\mathbf{P_s})$ and the experimental Helmholtz energy of solvation of the molecule s.

We carry out the derivation in two steps, and for convenience we use the T, V, N ensemble. Suppose first that s can attain only two conformations A and B, say the *cis* and *trans* conformations of a given molecule at equilibrium. The PCP of A is the change in the Helmholtz energy for placing an A molecule at a fixed position in l. The corresponding statistical mechanical expression is

$$\exp(-\beta\mu_A^{*l}) = \frac{q_A \int d\mathbf{X}^N d\mathbf{\Omega}_A \exp[-\beta U_N(\mathbf{X}^N) - \beta B_A(\mathbf{X}^N) - \beta U^*(A)]}{(8\pi^2) \int d\mathbf{X}^N \exp[-\beta U_N(\mathbf{X}^N)]} \tag{3.35}$$

where $B_A(\mathbf{X}^N)$ is the binding energy of A to the rest of the system of N molecules at configuration $\mathbf{X}^N$ (note that N is the sum of all molecules in the system including also any s molecules, but excluding only the solvaton, which is an A molecule); $U^*(A)$ denotes the intramolecular potential or the internal rotation potential function of s at the state A.

Clearly, integration over $\mathbf{\Omega}_A$ produces $8\pi^2$ and hence equation (3.35) may be rewritten as

$$\exp(-\beta\mu_A^{*l}) = q_A \exp[-\beta U^*(A)]\langle\exp(-\beta B_A)\rangle \tag{3.36}$$

where the average is over all configurations of the N molecules in the system. Likewise, for the gaseous phase we have

$$\exp(-\beta\mu_A^{*g}) = q_A \exp[-\beta U^*(A)] \tag{3.37}$$

Hence, the solvation Helmholtz energy of A is obtained from equations (3.36) and (3.37) in the form

$$\exp[-\beta\Delta\mu_A^{*l}] = \langle\exp(-\beta B_A)\rangle \tag{3.38}$$

and a similar expression holds true for B.

To obtain the connection between $\Delta\mu_A^{*l}$, $\Delta\mu_B^{*l}$, and $\Delta\mu_s^{*l}$, we start with the equilibrium condition

$$\mu_s^l = \mu_A^l = \mu_B^l \tag{3.39}$$

or, equivalently,

$$\mu_s^{*l} + kT\ln\rho_s^l \Lambda_s^3 = \mu_A^{*l} + kT\ln\rho_A^l \Lambda_s^3 = \mu_B^{*l} + kT\ln\rho_B^l \Lambda_s^3 \tag{3.40}$$

where $\rho_s^l = \rho_A^l + \rho_B^l$; ρ_A^l and ρ_B^l are the densities of A and B at equilibrium. Equation (3.40) may be rearranged to yield

$$\exp(-\beta\mu_s^{*l}) = \frac{\rho_s^l}{\rho_A^l}\exp(-\beta\mu_A^{*l}) \tag{3.41}$$

and

$$\exp(-\beta\mu_s^{*l}) = \frac{\rho_s^l}{\rho_B^l} \exp(-\beta\mu_B^{*l}) \tag{3.42}$$

On multiplying equation (3.41) by x_A^l and equation (3.42) by x_B^l, and adding the resulting two equations (where $x_A^l = \rho_A^l/\rho_s^l$ and $x_B^l = 1 - x_A^l$), we arrive at

$$\exp(-\beta\mu_s^{*l}) = \exp(-\beta\mu_A^{*l}) + \exp(-\beta\mu_B^{*l}) \tag{3.43}$$

This equation is equivalent to the statement that the partition function of a system with one additional s particle at a fixed position is the sum of the partition function of the same system with one A particle at a fixed position and the partition function of the same system with one B particle at a fixed position.

We now write the corresponding expression for the ideal-gas phase, namely

$$\exp(-\beta\mu_s^{*g}) = \exp(-\beta\mu_A^{*g}) + \exp(-\beta\mu_B^{*g}) \tag{3.44}$$

Taking the ratio between expressions (3.43) and (3.44) with the aid of equations (3.36) to (3.38), we obtain

$$\exp(-\beta\Delta\mu_s^{*l})$$
$$= \frac{q_A \exp[-\beta U^*(A)] \exp(-\beta\Delta\mu_A^{*l}) + q_B \exp[-\beta U^*(B)] \exp(-\beta\Delta\mu_B^{*l})}{q_A \exp[-\beta U^*(A)] + q_B \exp[-\beta U^*(B)]} \tag{3.45}$$

or equivalently

$$\exp(-\beta\Delta\mu_s^{*l}) = y_A^g \exp(-\beta\Delta\mu_A^{*l}) + y_B^g \exp(-\beta\Delta\mu_B^{*l}) \tag{3.46}$$

where y_A^g and y_B^g are the equilibrium mole fractions of A and B in the gaseous phase.

Generalization to the case with n discrete conformations is simply

$$\exp(-\beta\Delta\mu_s^{*l}) = \frac{\displaystyle\sum_{i=1}^{n} q_i \exp[-\beta U^*(i)]\langle\exp(-\beta B_i)\rangle}{\displaystyle\sum_{i} q_i \exp[-\beta U^*(i)]}$$

$$= \sum_{i=1}^{n} y_i^g \exp(-\beta\Delta\mu_i^{*l}) \tag{3.47}$$

Generalization to the continuous case is

$$\exp(-\beta \Delta\mu_s^{*l}) = \frac{\int d\mathbf{P}_s q(\mathbf{P}_s) \exp[-\beta U^*(\mathbf{P}_s)]\langle \exp[-\beta B(\mathbf{P}_s)]\rangle}{\int d\mathbf{P}_s \exp[-\beta U^*(\mathbf{P}_s)]}$$

$$= \int d\mathbf{P}_s y^g(\mathbf{P}_s) \exp[-\beta \Delta\mu^{*l}(\mathbf{P}_s)]$$

$$= \langle\langle \exp[-\beta B(\mathbf{P}_s)]\rangle\rangle \qquad (3.48)$$

where $q(\mathbf{P}_s)$ denotes the rotational, vibrational, etc., partition function of a single s molecule at a specific conformation $\mathbf{P}_s$, $y^g(\mathbf{P}_s)d\mathbf{P}_s$ is the mole fraction of s molecules at conformations between $\mathbf{P}_s$ and $\mathbf{P}_s + d\mathbf{P}_s$, and in the final expression on the rhs of relation (3.48) we have rewritten the integral as a double average quantity: one over all configurations of the N molecules (excluding the solvaton) and the second over all conformations of the s molecule with distribution function $y^g(\mathbf{P}_s)$.

In this section we have treated the T, V, N system. A similar treatment can be appied to the T, P, N system.

3.3. SOLVATION IN A TWO-PHASE SYSTEM AND MICELLAR SOLUTIONS

In the study of solubilization phenomena, it is often assumed that a micellar solution may be viewed as a two-phase system. Correspondingly, one then speaks of the solvation of a solute in each of the separate "phases." This point of view is not fully justified, however. In order to appreciate the difficulties in treating micellar solutions, we first discuss a simpler case: a two (real)-phase system.

Let A and B be two phases, and a component s be distributed between A and B and in equilibrium with a gaseous phase g. There are no restrictions on the compositions of the two phases, but we assume that g is an ideal gas.

The condition of equilibrium with respect to s is

$$\mu_s^g = \mu_s^l = \mu_s^A = \mu_s^B \qquad (3.49)$$

where μ_s^l is the CP of s in the liquid phase l, which consists of the phases A and B combined.

Equation (3.49) may now be rewritten as [using the general expression (3.11)]

$$\exp(-\beta\mu_s^{*l}) = \rho_s^l \exp(-\beta\mu_s^{*A})/\rho_s^A \tag{3.50}$$

$$\exp(-\beta\mu_s^{*l}) = \rho_s^l \exp(-\beta\mu_s^{*B})/\rho_s^B \tag{3.51}$$

where $\rho_s^A = N_s^A/V^A$, $\rho_s^B = N_s^B/V^B$, and $\rho_s^l = (N_s^A + N_s^B)/(V^A + V^B)$. Multiplying equation (3.50) by x_s^A and equation (3.51) by x_s^B, and adding the resulting two equations, we obtain

$$\exp(-\beta\mu_s^{*l}) = \eta^A \exp(-\beta\mu_s^{*A}) + \eta^B \exp(-\beta\mu_s^{*B}) \tag{3.52}$$

where $x_s^A = N_s^A/(N_s^A + N_s^B)$, $x_s^B = 1 - x_s^A$, $\eta^A = V^A/(V^A + V^B)$,   and $\eta^B = 1 - \eta^A$.

Dividing equation (3.52) by $\exp(-\beta\mu_s^{*g})$, we obtain the relation

$$\exp(-\beta\Delta\mu_s^{*l}) = \eta^A \exp(-\beta\Delta\mu_s^{*A}) + \eta^B \exp(-\beta\Delta\mu_s^{*B}) \tag{3.53}$$

This expression relates the solvation Gibbs energy of s in l, the combined liquid phases A and B, with the Gibbs energy of solvation of s in each phase separately. This equation is exact and contains well-defined quantities.

We now turn to aqueous micellar solutions. Suppose we can view a micellar solution as a quasi-two-phase system: the aqueous phase W and the micellar phase M. The question is, to what extent can we apply the relations derived above to this case? The main difficulty is that, in the micellar solutions, we do not have well-defined boundaries of the two quasi-phases. The very fact that we have written equation (3.49), (3.50), or (3.51) implicitly implies that we can imagine a partition that separates the two phases A and B. In fact, even the very definition of, say, μ_s^A implies that we can add one s to the phase A *only*, i.e., when we can introduce a (real or hypothetical) partition that permits s to wander in V^A and not in $V^A + V^B$. In writing equation (3.50), for instance, we have used the general relation for the CP, namely

$$\mu_s^A = \mu_s^{*A} + kT \ln \rho_s^A \Lambda_s^3 \tag{3.54}$$

which, according to the discussion in Section 3.1, may be interpreted in terms of a two-step process of adding one s to the phase A. First, we put an s molecule at a fixed position in A and then we release s to wander in the volume V^A. All of these considerations are clearly meaningful when A is a well-defined phase with a well-defined volume V^A.

The situation is markedly different when we treat micellar systems.

Obviously, we can write the CP of s in the entire micellar solution as

$$\mu_s^l = \mu_s^{*l} + kT \ln \rho_s^l \Lambda_s^3 \qquad (3.55)$$

where $\rho_s^l = N_s^l/V^l$, N_s^l being the *total* number of s molecules in the entire volume V^l. However, since we do not have well-defined (macroscopic) boundaries for the two quasi-phases, it is in principle impossible to write the analogue of equation (3.54) for the aqueous (W) and micellar (M) phases.

However, let us assume that we can define the boundaries of the micelles so that we can tell when s is in M or in W. Even if we could have devised a partition between the two phases, impermeable to s particles, we still face difficulties. Clearly, once we introduce such a boundary the corresponding CPs μ_s^M and μ_s^W become definable. What about the solvation Gibbs energy in M or in W? The answer is not simple.

Consider first the aqueous phase, and let us write the analogue of equation (3.54) for s in W:

$$\mu_s^W = \mu_s^{*W} + kT \ln \rho_s^W \Lambda_s^3 \qquad (3.56)$$

where μ_s^{*W} is the PCP of s in W. But what is ρ_s^W? If W is a real phase, then ρ_s^W must be N_s^W/V^W, where V^W is the volume of the phase W as defined by its boundaries. However, looking at the origin of the volume in expression (3.56) (see Section 3.1), we find that this should be the volume *accessible* to the solvaton molecule being released from a fixed position. Since the micelles *do* move about in the entire V^l, so does the freed s molecule. Hence, from this reasoning the density ρ_s^W should beN_s^W/V^l. Similarly, for the CP μ_s^{*M}, we could have written the formal analogue of equation (3.54) as

$$\mu_s^M = \mu_s^{*M} = kT \ln \rho_s^M \Lambda_s^3 \qquad (3.57)$$

Again, there is a difficulty in the very definition of the density ρ_s^M. Formally, it should have been $\rho_s^M = N_s^M/V^M$, but since the micelles move about in the entire volume V^l, so does the solvaton molecule that is included in the micelles.

All these difficulties arise because a micellar solution is not a real two-phase system. Therfore, the solvation Gibbs energy of s in each of the quasi-phases (or the more frequently discussed concept of the Gibbs energy of transferring s from W into M) is not a well-defined quantity, unless some drastic assumptions are made regarding the nature of the two phases.

3.4. ELEMENTS OF THE SCALED-PARTICLE THEORY (SPT)

In Section 2.2, we discussed the solvation process of a hard-sphere (HS) particle in a liquid consisting of HSs. At present there exists no exact theory that can predict the solvation Gibbs energy even for this simplest fluid. The SPT is perhaps the most successful theory aimed at that purpose. We therefore present here some of the fundamental ideas and results of this theory. The complete details of the theory are quite lengthy and complicated, and the interested reader should consult the original articles (for a review see Reiss[50]).

The germinal ideas of the SPT were first discussed by Hill,[51] who posed the problem of computing the cavity-size distribution in a fluid. More specifically, having a fluid consisting of N HSs at a given T and V, what is the probability of finding a cavity of radius r at some fixed position, say $\mathbf{R}_0$, in this system? A cavity is defined as a sphere of radius r centered at $\mathbf{R}_0$ from which all centers of the particles of the system are excluded. The SPT provides a recipe for computing an approximate value for this probability. This is equivalent to the computation of the solvation Hemholtz energy of a HS of suitable diameter.

Let us first introduce a very general relation between the Helmholtz energy change involved in the creation of a cavity of radius r at $\mathbf{R}_0$, and the probability of finding such a cavity in the fluid.

The Helmholtz energy change associated with the formation of a cavity may be written as

$$\exp[-\beta\Delta A(r)] = \exp[-\beta A(T, V, N; r) + \beta A(T, V, N)]$$

$$= \frac{\dfrac{1}{N!\,\Lambda^{3N}} \int \cdots_{V-v(r)} \int \exp[-\beta U(\mathbf{R}^N)]d\mathbf{R}^N}{\dfrac{1}{N!\,\Lambda^{3N}} \int \cdots_{V} \int \exp[-\beta U(\mathbf{R}^N)]d\mathbf{R}^N}$$

$$= P_0(r) \tag{3.58}$$

where the Helmholtz energy change $\Delta A(r)$ is expressed in terms of the ratio of two partition functions, corresponding to the same system (at T, V, N) with and without a cavity. The range of integration for the integral in the numerator is $V - v(r)$, i.e., the volume of the system minus the volume of the cavity $v(r) = 4\pi r^3/3$. The last equality on the rhs of relation (3.58) follows from the fundamental definition of the probability distribution in the T, V, N ensemble, i.e., the probability density of finding a specific configuration $\mathbf{R}^N = \mathbf{R}_1, \ldots, \mathbf{R}_N$ of the N particles is given by

$$P(\mathbf{R}^N) = \frac{\exp[-\beta U(\mathbf{R}^N)]}{\int \cdots_V \int \exp[-\beta U(\mathbf{R}^N)]d\mathbf{R}^N} \qquad (3.59)$$

Integrating equation (3.59) over all configurations $\mathbf{R}^N$ of the N particles, such that all the $\mathbf{R}_i$'s are to be excluded from the region $v(r)$, gives the probability of finding the system with such a region empty, i.e., with a spherical cavity of radius r.

We rewrite relation (3.58) in the form

$$P_0(r) = \int \cdots_{V-v(r)} \int P(\mathbf{R}^N)d\mathbf{R}^N$$

$$= \exp[-\beta \Delta A(r)] \qquad (3.60)$$

from which it follows that, since the integrand is always nonnegative, the integral is a monotonically decreasing function of r. In other words, the larger the size of the cavity, the smaller the probability of its spontaneous formation. This monotonic behavior was also noted in Section 2.2; however, here the statement is, of course, more general and independent of any assumption made in the SPT.

It was also noted in Section 2.2 that the creation of a cavity of radius r at $\mathbf{R}_0$ in a fluid of HSs with diameters σ_B is equivalent, from the standpoint of the thermodynamics of the system, to placing a HS particle of diameter σ_A at $\mathbf{R}_0$, such that

$$r = \frac{(\sigma_A + \sigma_B)}{2} \qquad (3.61)$$

Hence we have the following important identity between the two Helmoltz energy changes:

$$\Delta A_A^* = \Delta A(r) \qquad (3.62)$$

where ΔA_A^* is the Helmhotz energy of solvation of an A particle with diameter σ_A, in a fluid of B particles with diameters σ_B.

At this point we can derive two exact results for the function $\Delta A(r)$ in a system of HSs of diameter σ_B and density ρ. If the cavity is very small, such that $r \le \sigma_B/2$, it may be occupied by at most one particle at any given time. Therefore, the probability of finding the volume $v(r) =$

$4\pi r^3/3$ occupied is exactly $\rho v(r)$; hence the probability of finding the same volume empty is

$$P_0(r) = 1 - \rho 4\pi r^3/3 \qquad \text{for } r \leq \sigma_B/2 \qquad (3.63)$$

We note that a cavity of size $r = \sigma_B/2$ corresponds to a hypothetical HS particle of zero diameter $\sigma_A = 0$ [equation (3.61)].

On the other hand, if the cavity $v(r)$ becomes macroscopically large $(r \to \infty)$, the work required to build such a cavity is simply the thermodynamic work of compression, i.e.,

$$\Delta A(r) = Pv(r) \qquad \text{for } r \to \infty \qquad (3.64)$$

In equations (3.63) and (3.64) we have two exact results for the Helmholtz energy change required to form a cavity of very small and very large sizes. The SPT is basically a theory that bridges between these two exact values for $\Delta A(r)$. The starting point of the theory is the introduction of the function $P_0(r + dr)$, the probability that a cavity of radius $r + dr$ is found empty. (In the following we shall always refer to a cavity at some fixed position $\mathbf{R}_0$.) This function may be viewed as a product of two probabilities:

$$P_0(r + dr) = P_0(r)P_0(dr/r) \qquad (3.65)$$

where $P_0(r)$ was defined earlier, and $P_0(dr/r)$ is the conditional probability of finding the spherical shell of width dr empty, given that the sphere of radius r is empty. Equation (3.65), in fact, follows from the definition of the joint probability $P_0(r + dr)$ in terms of the conditional probability $P_0(dr/r)$.

Next, we introduce the auxiliary function $G(r)$ defined by

$$\rho 4\pi r^2 G(r)dr = 1 - P_0(dr/r) \qquad (3.66)$$

Clearly, since $P_0(dr/r)$ is the conditional probability of finding the shell "dr" empty, given that the cavity "r" is empty, the rhs of equation (3.66) is the conditional probability of finding at least one center of a particle in the shell "dr", given that the cavity "r" is empty. The function $G(r)$ was found useful for the construction of the approximate bridge between the two extreme results (3.63) and (3.64).

If we expand $P_0(r + dr)$ to first order in dr, we obtain

$$P_0(r + dr) = P_0(r) + \frac{dP_0(r)}{dr} dr \qquad (3.67)$$

Hence using equations (3.65) and (3.66) we have

$$\frac{d \ln P_0(r)}{dr} = -4\pi r^2 \rho G(r) \tag{3.68}$$

which, upon integration, yields

$$\ln P_0(r) - \ln P_0(r = 0) = -\rho \int_0^r 4\pi \lambda^2 G(\lambda) d\lambda \tag{3.69}$$

Since the probability of finding a cavity of radius $r = 0$ is unity [this follows from the definition (3.60)], equation (3.69) may be reduced to

$$\ln P_0(r) = -\rho \int_0^r 4\pi \lambda^2 G(\lambda) d\lambda \tag{3.70}$$

or, equivalently,

$$\Delta A^*(r) = -kT \ln P_0(r)$$

$$= kT\rho \int_0^r 4\pi \lambda^2 G(\lambda) d\lambda \tag{3.71}$$

Here, we express the work required to create a cavity of radius r as a "scaling process," i.e., we build up the cavity from $\lambda = 0$ to $\lambda = r$ by a continuous process. This is also equivalent to the process of building up a HS particle at some fixed position in the system: hence the origin of the name *scaled-particle* theory.

Reiss *et al.*[50,52,53] suggested a simple analytical form for the function $G(r)$, namely

$$G(r) = A + Br^{-1} + Cr^{-2} \tag{3.72}$$

where the coefficients A, B, and C are determined by using all the available information on the behavior of the function $G(r)$. In particular, we have the two limiting forms of behavior, which follow from equations (3.63) and (3.64), namely

$$G(r) = (1 - 4\pi r^3 \rho/3)^{-1} \qquad \text{for } r \leq \sigma_B/2 \tag{3.73}$$

$$G(r) = P/kT\rho \qquad \text{for } r \to \infty \tag{3.74}$$

Once the three coefficients A, B, and C are determined, one may proceed to integrate equation (3.71) to obtain $\Delta A^*(r)$. The final result is

$$\Delta A^*(r) = K_0 + K_1 r + K_2 r^2 + K_3 r^3 \tag{3.75}$$

with the coefficients K_i given by

$$K_0 = kT[-\ln(1-y) + 4.5z^2] + \pi P \sigma_B^3/6 \qquad (3.76)$$

$$K_1 = -(kT/\sigma_B)(6z + 18z^2) + \pi P \sigma_B^2 \qquad (3.77)$$

$$K_2 = (kT/\sigma_B^2)(12z + 18z^2) - 2\pi P \sigma_B \qquad (3.78)$$

$$K_3 = 4\pi P/3 \qquad (3.79)$$

and

$$y = \rho \pi \sigma_B^3/6, \qquad z = y/(1-y) \qquad (3.80)$$

Thus, in essence the SPT provides an expression for the work required to create a cavity of radius r at $\mathbf{R}_0$. This is equivalent to the solvation Helmholtz energy of one A particle of diameter σ_A in a fluid of particles with diameters σ_B, such that relation (3.61) is fulfilled. It is noteworthy that in this section we have been discussing the solvation Helmholtz energy in a T, V, N system. However, as discussed in Section 3.1, this is the same as the Gibbs energy of solvation in a T, P, N system provided the average volume $\langle V \rangle$ in the latter is equal to the exact volume V in the former system.

We conclude this section with some general comments regarding the application of the SPT to real systems. In the first place, as was also emphasized in Section 2.2, the identity between the Gibbs energy of a cavity formation and solvation Gibbs energy of a HS (of suitable diameter) holds only when the cavity is created at a *fixed* point, and only when we use the term *solvation* as defined in Section 1.2. There exists no definition of a "chemical potential of a cavity." It therefore follows that one cannot define the conventional "standard Gibbs energy of solvation of a cavity." Yet these concepts do appear quite frequently in the literature.

Second, the SPT was originally devised for HS fluids, where the particles are characterized by one parameter only — their diameter.[†] The remarkable success of the theory for other systems, such as inert-gas liquids, aqueous solutions, and ionic fluids, is quite surprising. However, one must bear in mind that the SPT is not a pure molecular theory in the sense that it does not attempt to predict the thermodynamic properties of a system from knowledge of its molecular parameters alone. An ideal molecular theory should furnish, say, the Gibbs energy as a function of T and P and all the molecular parameters $a_1, \ldots, a_n$, i.e., a function $G(T, P; a_1, \ldots, a_n)$. Instead, the SPT replaces all of the molecular

---

[†] This is true in the context of our interest in solvation thermodynamics. For a full description of a system of HSs, one needs both the diameter and mass of the particles.

parameters $a_1, \ldots, a_n$ by one parameter, the effective diameter of the molecule. This deficiency in the input information is compensated by the injection of the measurable density into the theory (a quantity that, in principle, should be calculable by a molecular theory). This procedure accounts for the remarkable success of the theory, yet it is also its main weakness. Clearly, the theory cannot provide us with any insight into the molecular or structural aspects of the macroscopically observable quantities.

3.5. CONDITIONAL SOLVATION GIBBS ENERGY; HARD AND SOFT PARTS

One of the earliest attempts to interpret the anomalous properties of aqueous solutions was made by Eley,[25,26] who, following some earlier ideas, suggested splitting the solvation process into two parts: a cavity formation and introduction of the solute into the cavity. In the present section we shall discuss the statistical mechanical basis for such a split of the solvation process.

In essence, the molecular basis for splitting the solvation into two (or more) steps stems from the recognition of the two (or more) parts of the solute–solvent intermolecular potential function, the so-called hard and soft parts of the interactions.

For simplicity let us assume that the solute–solvent pair potential is a function of the distance R only, and that this function may be written as

$$U_{sb}(R) = U_{sb}^H(R) + U_{sb}^S(R) \qquad (3.81)$$

where the "hard" part of the potential is defined through a choice of effective hard-core diameter for the solute σ_s and solvent σ_b, respectively, namely

$$U_{sb}^H(R) = \begin{cases} \infty & \text{for } R \le \sigma_{sb} \\ 0 & \text{for } R > \sigma_{sb} \end{cases} \qquad (3.82)$$

where $\sigma_{sb} = (\sigma_s + \sigma_b)/2$. One can always find a small enough value of σ_{sb} such that, for $R \le \sigma_{sb}$, the potential function rises so steeply as to justify an approximation of the form (3.82). The "soft" part, U_{sb}^S, is next defined through equation (3.81).

Note that we have used here the terms solute and solvent in their conventional sense. The same procedure may be employed for a pure liquid. However, when we have a mixture of several components at an arbitrary composition, there exists no well-defined concept of a cavity suitable to accommodate the solvaton (see also Section 2.2).

Except for HS particles, the split of the potential function into two terms, as in equation (3.81), always contains an element of ambiguity. This ambiguity is propagated into the split of the solvation Gibbs energy, to be carried out below.

We now write the solvation Gibbs energy of s as

$$\Delta G_s^* = -kT \ln\langle \exp(-\beta B_s)\rangle$$
$$= -kT \ln\langle \exp(-\beta B_s^H - \beta B_s^S)\rangle \tag{3.83}$$

where B_s^H and B_s^S are the "hard" and "soft" parts of the binding energies of s to all the solvent molecules.[†]

The average quantity in relation (3.83) can be viewed as an average of a product of two functions. If these were independent (in the sense of independence of two random variables), one could rewrite the average as a product of two averages, i.e.,

$$\langle \exp(-\beta B_s^H)\exp(-\beta B_s^S)\rangle = \langle \exp(-\beta B_s^H)\rangle\langle \exp(-\beta B_s^S)\rangle \tag{3.84}$$

and ΔG_s^* could be split into the corresponding factors:

$$\Delta G_s^* = \Delta G_s^{*H} + \Delta G_s^{*S} \tag{3.85}$$

However, the factorization in equation (3.84) is in general invalid, and therefore a split of ΔG_s^* in terms of a hard and a soft part, in the form of (3.85), may not be justified.

Instead, we may rewrite the average in relation (3.83) as

$$\langle \exp(-\beta B_s^H)\exp(-\beta B_s^S)\rangle$$

$$= \frac{\int d\mathbf{X}^N \exp(-\beta U_N - \beta B_s^H)\exp(-\beta B_s^S)}{\int d\mathbf{X}^N \exp(-\beta U_N)}$$

$$= \frac{\int d\mathbf{X}^N \exp(-\beta U_N - \beta B_s^H)}{\int d\mathbf{X}^N \exp(-\beta U_N)}\frac{\int d\mathbf{X}^N \exp(-\beta U_N - \beta B_s^H)\exp(-\beta B_s^S)}{\int d\mathbf{X}^N \exp(-\beta U_N - \beta B_s^H)}$$

$$= \langle \exp(-\beta B_s^H)\rangle\langle \exp(-\beta B_s^S)\rangle_H \tag{3.86}$$

[†] It is here that we assume that the same σ_{sb} is chosen for all the U_{si}^H. This clearly holds for s in pure b, or s in pure s. If the solvent is a mixture, then it is less clear how to define a general cutoff for the potentials U_{si}.

where the second average on the rhs of equation (3.86) is a *conditional* average, using the probability distribution function

$$P(\mathbf{X}^N/\mathbf{X}_s^H) = \frac{P(\mathbf{X}^N, \mathbf{X}_s^H)}{P(\mathbf{X}_s^H)} = \frac{\exp(-\beta U_N - \beta B_s^H)}{\int d\mathbf{X}^N \exp(-\beta U_N - \beta B_s^H)} \tag{3.87}$$

with $P(\mathbf{X}^N/\mathbf{X}_s^H)$ the probability density of finding a configuration $\mathbf{X}^N$ of the N particles given a hard particle at $\mathbf{X}_s^H$. Thus, with the aid of equation (3.87), the solvation Gibbs energy may be split as

$$\Delta G_s^* = \Delta G_s^{*H} + \Delta G_s^{*S/H} \tag{3.88}$$

where ΔG_s^{*H} is the Gibbs energy of solvation of the hard part of the potential (this corresponds to the Gibbs energy of cavity formation in a simple system of HS particles); $\Delta G_s^{*S/H}$ is the *conditional* Gibbs energy of solvation of the soft part of the potential. It is the work required to couple, or to "switch on," the soft part of the solute–solvent interaction, given that the hard part of the potential has already been solvated.

In the original papers of Eley[25,26] and in some more recent publications, one finds an analogue of equation (3.85) in terms of the conventional standard Gibbs energy of solvation (or of solution) in the form

$$\Delta G_s^\circ = \Delta G_{s,\,cavity}^\circ + \Delta G_{s,\,interaction}^\circ \tag{3.89}$$

However, as noted in Section 3.4, ΔG_s° is defined through the CP of the solute s. Since there exists no definition of the CP of a cavity, one cannot construct the quantity $\Delta G_{s,\,cavity}^\circ$ from the purely conventional approach. One should use either a HS solute or the solvation concept as defined in this book to effect the split of ΔG_s° as in expression (3.88).

Finally, we note that the procedure outlined in this section may be generalized to include different contributions to the pair potential. An important generalization of equation (3.81) might be the inclusion of electrostatic interactions, e.g., charge–dipole interaction between an ionic solute and the dipole moment of a solvent molecule. The generalization of equation (3.81) might look like this:

$$U_{sb}(R) = U_{sb}^H(R) + U_{sb}^S(R) + U_{sb}^{el}(R) \tag{3.90}$$

Correspondingly, the solvation Gibbs energy of s will be written, in a generalization of expression (3.88), as follows:

$$\Delta G_s^* = \Delta G_s^{*H} + \Delta G_s^{*S/H} + \Delta G_s^{*el/S,H} \tag{3.91}$$

where $\Delta G_s^{*el/S,H}$ is the *conditional* Gibbs energy of solvation of the electrostatic part of the interaction, given that a solvaton with the soft and hard parts (excluding the electrostatic part) has already been placed at a fixed position in the solvent.

It is important to recognize the conditional character of $\Delta G_s^{*el/S,H}$, since the charging work of a "free" solute might be considerably different from that for a solute at a fixed position.

3.6. CONDITIONAL SOLVATION GIBBS ENERGY; GROUP ADDITIVITY

Researchers in the field of solvation thermodynamics have long attempted to assign group contributions of various parts of a solute molecule to the thermodynamic quantities of solvation. In this section we examine the molecular basis of such a group-additivity approach. As we shall see, the problem is essentially the same as that treated in Section 3.5, i.e., it originates from a split of the solute–solvent intermolecular potential function into two or more parts.

Let us start with a solute of the form X–Y, where X and Y are two groups, say CH_3 and CH_3 in ethane or CH_3 and OH in methanol. We assume that the solute–solvent interaction may be split into two parts as follows:

$$U(X-Y, i) = U(X, i) + U(Y, i) \qquad (3.92)$$

where $U(X, i)$ and $U(Y, i)$ are the interaction energies between the groups X and Y and the ith solvent molecule, respectively.

As in the case of Section 3.5, where we had split the interaction energy into a hard and a soft part, we also have here an element of ambiguity as to the exact manner in which this split may be achieved. However, assuming that an equation of the form (3.92) may be applied, we can proceed to split the total binding energy of the solute X–Y into two parts, simply by summing the interactions between the solute and all solvent molecules at some particular configuration, say $\mathbf{X}^N$; thus

$$B_{X-Y} = \sum_{i=1}^{N} U(X, i) + \sum_{i=1}^{N} U(Y, i)$$
$$= B_X + B_Y \qquad (3.93)$$

The solvation Gibbs energy of the solute is now written as

$$\Delta G_{X-Y}^* = -kT \ln\langle \exp(-\beta B_{X-Y})\rangle$$
$$= -kT \ln\langle \exp(-\beta B_X) \exp(-\beta B_Y)\rangle \qquad (3.94)$$

We see that, as in Section 3.5, we have here an average of a product of two functions. This, in general, may not be factorized into a product of two average quantities. If this could have been done, then relation (3.94) could have been written as a sum of two terms, i.e.,

$$\Delta G^*_{X-Y} = - kT \ln \langle \exp(-\beta B_X) \rangle - kT \ln \langle \exp(-\beta B_Y) \rangle$$
$$= \Delta G^*_X + \Delta G^*_Y \tag{3.95}$$

Such a group additivity, though very frequently assumed to hold for ΔG^*_{X-Y} (as well as for other thermodynamic quantities), has clearly no molecular basis, even if we assume that a split of the potential function in expression (3.92) is exact. It is noteworthy that the quantity $\Delta G^*_{X-Y} - \Delta G^*_X - \Delta G^*_Y$ [presumed to be zero in relation (3.95)] is exactly equal to the indirect part of the Gibbs energy change for the process of bringing the two groups X and Y from infinite separation into the final configuration as in the molecule X–Y. This aspect is further discussed in connection with the hydrophobic-interaction problem in Section 3.18.

In order to achieve some form of additivity similar to relation (3.95), we rewrite equation (3.94) as follows:

$$\Delta G^*_{X-Y} = - kT \ln \langle \exp(-\beta B_X - \beta B_Y) \rangle$$

$$= - kT \ln \left[\frac{\int d\mathbf{X}^N \exp(-\beta U_N) \exp(-\beta B_X) \exp(-\beta B_Y)}{\int d\mathbf{X}^N \exp(-\beta U_N) \exp(-\beta B_X)} \right.$$

$$\left. \times \frac{\int d\mathbf{X}_N \exp(-\beta U_N) \exp(-\beta B_X)}{\int d\mathbf{X}^N \exp(-\beta U_N)} \right]$$

$$= - kT \ln \left[\int d\mathbf{X}^N P(\mathbf{X}^N / \mathbf{X}_X) \exp(-\beta B_Y) \int d\mathbf{X}^N P(\mathbf{X}^N) \exp(-\beta B_X) \right]$$

$$= \Delta G^*_{Y/X} + \Delta G^*_X \tag{3.96}$$

The significance of equation (3.96) is the following. The solvation Gibbs energy of the solute X–Y is written [exactly, presuming the validity of relation (3.93)] as a sum of two contributions: first, the solvation Gibbs energy of the group X, and second, the *conditional* solvation Gibbs energy of the group Y at an adjacent point near the group X given that the group X is already at a fixed point in the liquid.

Clearly, since the second group, Y, is brought to a location very near to X, one cannot ignore the effect of X on the (conditional) solvation Gibbs energy of Y. It is only in the very extreme cases when X and Y are very far apart that we could assume that their solvation Gibbs energies will be strictly additive, as in equation (3.95).

Because of symmetry we could, of course, replace equation (3.96) by the equivalent relation

$$\Delta G^*_{X-Y} = \Delta G^*_{X/Y} + \Delta G^*_Y \qquad (3.97)$$

The procedure outlined above may evidently be generalized to include large molecules with many groups. A rigorous split in the form (3.95) cannot be expected in general, but a generalized form of relation (3.96) or (3.97) can be derived based on the concept of the conditional solvation Gibbs energy.

As we have noted in Section 3.5, an additivity of the form (3.95) is frequently applied to the *conventional* standard Gibbs energy of solvation. The latter sometimes contains residues of the liberation Gibbs energy, a part which cannot be split into group contributions in an obvious way.

3.7. DEFINITION OF THE STRUCTURE OF WATER

The concept of the structure of water (SOW) has been used in the physical-chemical literature ever since scientists were concerned with the peculiar properties of liquid water and aqueous solutions. In this section we present one possible definition of the SOW which will serve us in the following few sections. This definition is based on a particular choice of the water–water pair potential. We believe that the quantity we shall be referring to as the SOW is in conformity with current ideas and views on the SOW as has been expressed in many articles and books dealing with this topic.

For convenience let us start with the assumption that the water–water pair potential has the general form (more details are given in Chapter 6 of Ben-Naim[4])

$$U(\mathbf{X}_i, \mathbf{X}_j) = U(R_{ij}) + U_{el}(\mathbf{X}_i, \mathbf{X}_j) + \varepsilon_{HB}G(\mathbf{X}_i, \mathbf{X}_j) \qquad (3.98)$$

where $U(R_{ij})$ is a spherically symmetric contribution to the total interaction between two water molecules. This part may conveniently be chosen to have the Lennard-Jones form, with the appropriate parameters of neon, an atom isoelectronic with a water molecule. Its main function is to account for the very short-range interaction. The electrostatic interaction part $U_{el}(\mathbf{X}_i, \mathbf{X}_j)$ may include the interactions between a few electric multipoles, such as dipole and quadrupoles. This part is to account for the long-range interaction between two water molecules. The third term on the rhs of equation (3.98) is to account for the interaction energy at intermediate distances around 2.8 Å. There exists no information on the analytical form of the potential function in this intermediate range of distances. However, recognizing the fact that two water molecules form a hydrogen bond (HB) at some quite well-defined configuration $(\mathbf{X}_i, \mathbf{X}_j)$, we shall refer to this part of the potential as the HB part. In the literature one can find several suggestions for an explicit form for this part of the potential that was used to simulate the properties of liquid water and aqueous solutions. However, for all our purposes we shall not need to describe this function in any detail. Instead, it will be sufficient to describe one important feature of this function that is essential for our definition of the SOW.

We assume that the HB part consists of an energy parameter ε_{HB}, which we refer to as the HB energy, and a geometric function $G(\mathbf{X}_i, \mathbf{X}_j)$, which is essentially a stipulation on the relative configuration of the pair of molecules i and j. Namely, this function attains a maximum value of unity whenever the configuration of the two molecules is most favorable to form a HB. Its value drops sharply to zero when the configuration deviates considerably from the one required for the formation of a HB.

With this qualitative description we define the function $G(\mathbf{X}_i, \mathbf{X}_j)$ by

$$G(\mathbf{X}_i, \mathbf{X}_j) = \begin{cases} 1 & \text{if } i \text{ and } j \text{ are in configurations} \\ & \text{favorable for a HB formation} \\ 0 & \text{for all other configurations} \end{cases} \tag{3.99}$$

Although we did not specify which configurations are favorable for a HB formation, we have in mind configurations such that the O–O distance is about 2.76 Å and that one of the O–H bonds on one molecule is directed toward one of the lone pair of electrons on the second moecule. Also, in equation (3.99) we let the value of $G(\mathbf{X}_i, \mathbf{X}_j)$ drop sharply to zero. One way of achieving such a drop in a continuous manner is by using Gaussian functions or similar functions (for more details, see Chapter 6 in Ben-Naim[4]).

Clearly, with the above qualitative description of $G(\mathbf{X}_i, \mathbf{X}_j)$, we leave a great deal of freedom as to the manner in which we split the potential (3.98) into its various contributions, and as to the specific form we wish to choose for this function. The important feature that we need for our definition of the SOW is the following: let $\mathbf{X}^N = \mathbf{X}_1, \dots, \mathbf{X}_N$ be a specific configuration of all the N water molecules in the system at some T and P.

We select one molecule, say the ith one, and define the quantity

$$\psi_i(\mathbf{X}^N) = \sum_{\substack{j=1 \\ j \neq i}}^{N} G(\mathbf{X}_i, \mathbf{X}_j) \tag{3.100}$$

Since $G(\mathbf{X}_i, \mathbf{X}_j)$ contributes unity to the sum on the rhs of equation (3.100) whenever the jth molecule is hydrogen bonded to the ith molecule, $\psi_i(\mathbf{X}^N)$ measures the number of HBs in which the ith molecule participates when the entire system is at the configuration $\mathbf{X}^N$. Based on what we know about the behavior of water molecules, $\psi_i(\mathbf{X}^N)$ may attain values between zero and four.

If $\psi_i(\mathbf{X}^N) = 4$, we may say that the local structure around the ith molecule, at the configuration $\mathbf{X}^N$, is similar to that of ice. If $\psi_i(\mathbf{X}^N) = 0$, the local structure around the ith molecule bears no resemblance to that of ice. Hence we can use $\psi_i(\mathbf{X}^N)$ to serve as a measure of the *local* structure around the ith molecule at the given configuration of the whole system.

Next, we define the average of $\psi_i(\mathbf{X}^N)$ in the T, P, N ensemble:

$$\langle \psi \rangle_0 = \int dV \int d\mathbf{X}^N P(\mathbf{X}^N, V) \psi_i(\mathbf{X}^N) \tag{3.101}$$

where $P(\mathbf{X}^N, V)$ is the fundamental distribution function in the T, P, N ensemble. Since all molecules in the system are equivalent, the average in expression (3.101) is independent of the subscript i; hence we denote this quantity by $\langle \psi \rangle_0$, and not $\langle \psi_i \rangle_0$. The subscript zero indicates that the average is taken with the distribution $P(X, V)$ for *pure* water; this should be distinguished from a conditional average discussed in Sections 3.8 and 3.9.

Thus $\langle \psi \rangle_0$ is the average number of HBs formed by (any) single water molecule in pure water at a given T and P. This quantity may serve as a definition of the local structure around a given molecule in the system.

The total average number of HBs in the entire system is evidently

$$\langle \text{HB} \rangle_0 = \frac{N}{2} \langle \psi \rangle_0 \tag{3.102}$$

The division by 2 is required, since in $N\langle \psi \rangle_0$ we count each HB twice.

From the definition of $\psi_i(\mathbf{X}^N)$, we may also rewrite equation (3.102) in the form

$$
\langle HB \rangle_0 = \frac{N}{2} \int dV \int dX^N P(X^N, V) \sum_{\substack{j=1 \\ j \neq i}}^{N} G(X_i, X_j)
$$

$$
= \frac{1}{2} \int dV \int dX^N P(X^N, V) \sum_{i=1}^{N} \sum_{\substack{j=1 \\ j \neq i}}^{N} G(X_i, X_j)
$$

$$
= \int dV \int dX^N P(X^N, V) \sum_{\substack{i=1 \\ i<j}}^{N} \sum_{j=1}^{N} G(X_i, X_j) \qquad (3.103)
$$

In subsequent sections we shall use the last form of the rhs of relation (3.103) as our definition of the structure of water. Clearly, either $\langle \psi \rangle_0$ or $\langle HB \rangle_0$ can be used for that purpose. Also, as we shall see in the next sections, we need not assume a full pairwise additivity of the total potential energy of the system. Instead, we require that

$$
U(X^N) = U'(X^N) + \varepsilon_{HB} \sum_{\substack{i=1 \\ i<j}}^{N} \sum_{j=1}^{N} G(X_i, X_j) \qquad (3.104)
$$

where in U' we lumped together both the Lennard-Jones and the electrostatic interactions among all the N water molecules. The pairwise additivity is required only from the total HB interactions, as is explicitly written on the rhs of equation (3.104).

3.8. FORMAL RELATIONS BETWEEN SOLVATION THERMODYNAMICS AND THE STRUCTURE OF WATER

We consider first the case of a simple solute s in a very dilute solution in water w. By very dilute we mean that all solute–solute interactions may be neglected. Formally, this is equivalent to a system containing just one s solute and N water molecules. Also, for convenience we assume that the system is at a given temperature T and volume V. The Helmholtz energy of solvation of s in this system is

$$
\Delta A_s^* = -kT \ln \langle \exp(-\beta B_s) \rangle_0 \qquad (3.105)
$$

where

$$\langle \exp(-\beta B_s) \rangle_0 = \frac{\int d\mathbf{X}^N \exp[-\beta U_N(\mathbf{X}^N) - \beta B_s(\mathbf{R}_s, \mathbf{X}^N)]}{\int d\mathbf{X}^N \exp[-\beta U_N(\mathbf{X}^N)]}$$

$$= \int d\mathbf{X}^N P(\mathbf{X}^N) \exp[-\beta B_s(\mathbf{R}_s, \mathbf{X}^N)] \qquad (3.106)$$

We have denoted by $B_s(\mathbf{R}_s, \mathbf{X}^N)$ the total binding energy of s to all the N water molecules at a specific configuration $\mathbf{X}^N$. The solute s is presumed to be at some fixed position $\mathbf{R}_s$; however, since the choice of $\mathbf{R}_s$ is irrelevant to the value of the solvation Helmholtz energy, it does not appear in the notation on the lhs of equation (3.106). We note also that the average $\langle \ \rangle_0$ is taken with the probability distribution of the N water molecules of *pure* liquid water, i.e., without the presence of a solute s at $\mathbf{R}_s$. The subscript zero serves to distinguish this average from a conditional average introduced below.

The solvation entropy of s is obtained by taking the derivative of equation (3.105) with respect to T, i.e.,

$$\Delta S_s^* = -(\partial \Delta A_s^*/\partial T)_{V,N}$$

$$= k \ln\langle \exp(-\beta B_s)\rangle_0 + \frac{1}{T}(\langle B_s \rangle_s + \langle U_N \rangle_s - \langle U_N \rangle_0) \quad (3.107)$$

where the symbol $\langle \ \rangle_s$ signifies a *conditional* average over all configurations of the solvent molecules, given a solute s at a fixed position $\mathbf{R}_s$; that is, we use the conditional distribution defined by

$$P(\mathbf{X}^N/\mathbf{R}_s) = \frac{\exp[-\beta U_N(\mathbf{X}^N) - \beta B_s(\mathbf{R}_s, \mathbf{X}^N)]}{\int d\mathbf{X}^N \exp[-\beta U_N(\mathbf{X}^N) - \beta B_s(\mathbf{R}_s, \mathbf{X}^N)]} \qquad (3.108)$$

The solvation energy is obtained from equations (3.105) and (3.107) in the form

$$\Delta E_s^* = \Delta A_s^* + T\Delta S_s^* = \langle B_s \rangle_s + \langle U_N \rangle_s - \langle U_N \rangle_0 \qquad (3.109)$$

As noted in Section 3.21, for all the systems of interest in this book there is not much difference between the values of ΔG_s^* and ΔA_s^* or between ΔH_s^* and ΔE_s^*.

The solvation energy as presented in equation (3.109) has a very simple interpretation. It consists of an average binding energy of s to the system, and a change in the average total interaction energy among all the N water molecules, induced by the solvation process. For any solvent, this change in the total potential energy may be reinterpreted as a structural change induced by s on the solvent (this aspect is treated in great detail in Chapter 5 of Ben-Naim[4]). For the special case of liquid water, we use the split of the total potential energy as in equation (3.104) to rewrite equation (3.109) as

$$\Delta E_s^* = \langle B_s \rangle_s + \langle U_N' \rangle_s - \langle U_N' \rangle_0 + \varepsilon_{HB}(\langle HB \rangle_s - \langle HB \rangle_0) \quad (3.110)$$

Hence we see that ΔE_s^* contains explicitly a term that originates from the change in the average number of HBs in the system, induced by the solvation process. We also note that the same term appears also in the solvation entropy as written in equation (3.107). Furthermore, when we form the combination of $\Delta E_s^* - T\Delta S_s^*$, this term cancels out. The conclusion is that any structural change in the solvent induced by the solute might affect ΔE_s^* and ΔS_s^* but will have no effect on the solvation Helmholtz energy. This conclusion has been derived in earlier publications[30,54] for more general processes and for more general notions of structural changes in the solvent.

The second case of interest is the solvation of a water molecule in pure liquid water. Again we assume that we have a T, V, N system, and we add one water molecule to a fixed position, say $\mathbf{R}_w$. The solvation Helmholtz energy is

$$\Delta A_w^* = -kT \ln \langle \exp(-\beta B_w) \rangle_0 \quad (3.111)$$

where $\langle \ \rangle_0$ indicates an average over all the configurations of the N water molecules excluding the solvaton. The corresponding entropy and energy of solvation of a water molecule are obtained by standard relationships. The results are

$$\Delta S_w^* = k \ln \langle \exp(-\beta B_w) \rangle_0 + \frac{1}{T}(\langle B_w \rangle_w + \langle U_N \rangle_w - \langle U_N \rangle_0) \quad (3.112)$$

and

$$\Delta E_w^* = \langle B_w \rangle_w + \langle U_N \rangle_w - U_N \rangle_0 \quad (3.113)$$

where here the conditional average is taken with the probability distribution

$$P(\mathbf{X}^N/\mathbf{X}_w) = \frac{\exp[-\beta U_{N+1}(\mathbf{X}^{N+1})]}{\displaystyle\int d\mathbf{X}^N \exp[-\beta U_{N+1}(\mathbf{X}^{N+1})]} \tag{3.114}$$

which is the probability density of finding a configuration $\mathbf{X}^N$ given one water molecule at a specific configuration $\mathbf{X}_w$. We note that fixing the configuration $\mathbf{X}_w$ of one of the water molecules is the same as fixing only the location of this molecule. This is so, since we can perform the integration over all orientations of this molecule and a factor $8\pi^2$ will cancel out:

$$P(\mathbf{X}^N/\mathbf{X}_w) = \frac{P(\mathbf{X}^N, \mathbf{X}_w)}{P(\mathbf{X}_w)} = \frac{P(\mathbf{X}^N, \mathbf{R}_w)}{P(\mathbf{R}_w)} = P(\mathbf{X}^N/\mathbf{R}_w) \tag{3.115}$$

Formally, equations (3.112) and (3.113) are similar to the corresponding equations (3.107) and (3.109). It follows that one can give also a formal interpretation to the various terms as we have done before. However, since here we have the case of *pure* liquid water, it is clear that the average binding energy of a water molecule is the same as the *conditional* average binding energy of a water molecule, i.e.,

$$\langle B_w \rangle_w = \int d\mathbf{X}^N P(\mathbf{X}^N/\mathbf{X}_w) B_w = \frac{\displaystyle\int d\mathbf{X}^N P(\mathbf{X}^N, \mathbf{X}_w) B_w}{P(\mathbf{X}_w)}$$

$$= \frac{\displaystyle\int d\mathbf{X}^N P(\mathbf{X}^N, \mathbf{X}_w) B_w}{8\pi^2/V} = \int d\mathbf{X}_w d\mathbf{X}^N P(\mathbf{X}^N, \mathbf{X}_w) B_w = \langle B_w \rangle_0 \tag{3.116}$$

where $\langle B_w \rangle_0$ is the average binding energy of a water molecule in a system of pure water (of either N or $N + 1$ molecules). Furthermore

$$\langle U_N \rangle_w = \frac{\displaystyle\int d\mathbf{X}^N P(\mathbf{X}^N, \mathbf{X}_w) U_N}{P(\mathbf{X}_w)} = \frac{\displaystyle\int d\mathbf{X}^N P(\mathbf{X}^N, \mathbf{X}_w) U_N}{8\pi^2/V}$$

$$= \int d\mathbf{X}_w d\mathbf{X}^N P(\mathbf{X}^N, \mathbf{X}_w) U_N = \int d\mathbf{X}_w d\mathbf{X}^N P(\mathbf{X}^N, \mathbf{X}_w)(U_{N+1} - B_w)$$

$$= \langle U_{N+1} \rangle_0 - \langle B_w \rangle_0 \tag{3.117}$$

Using the latter equation we can rewrite the expression for ΔE_w^* [equation (3.113)] in the simpler form

$$\Delta E_w^* = \langle B_w \rangle_0 + \frac{N+1}{2} \langle B_w \rangle_0 - \langle B_w \rangle_0 - \frac{N}{2} \langle B_w \rangle_0$$

$$= \tfrac{1}{2}\langle B_w \rangle_0 \qquad\qquad (3.118)$$

Thus the solvation energy of a water molecule in pure liquid water is simply half the average binding energy of a single water molecule. This result is of course more general, and applies to any liquid.[4]

Finally, we note that for convenience we have used the T, V, N ensemble in this section. Similar results may be obtained in any other ensemble as well.

3.9. ISOTOPE EFFECT ON SOLVATION THERMODYNAMICS AND STRUCTURAL ASPECTS OF AQUEOUS SOLUTIONS

In Section 3.8 we derived a general and formal relationship between thermodynamics of solvation and structural changes induced in the water. We now present an approximate relationship between the structure of water and the isotope effect on the solvation Gibbs energy of a water molecule. Similarly, from the isotope effect on the solvation Gibbs energy of an inert solute, we can estimate the extent of structural change induced by the solute on the solvent. The presentation in this section is quite brief and more detailed treatment may be found elsewhere (see Chapter 5 of Ben-Naim,[30] and elsewhere[31,32]).

(1) *Pure liquid water*. As in Section 3.7, we assume in this section that water molecules interact according to a potential function of the form (3.98). We also assume that H_2O and D_2O have essentially the same potential function, except for the HB energy parameter. It is assumed that $|\varepsilon_{HB}|$ is slightly larger for D_2O than for H_2O. We denote by ε_D and ε_H the energy parameters for D_2O and H_2O, respectively.

Viewing ΔG_w^* as a function of the HB energy parameter, we expand $\Delta G_{D_2O}^*$ about $\Delta G_{H_2O}^*$ to first order in ε_{HB} and obtain

$$\Delta G_{D_2O}^* = \Delta G_{H_2O}^* + \frac{\partial \Delta G_w^*}{\partial \varepsilon_{HB}} (\varepsilon_D - \varepsilon_H) + \cdots \qquad (3.119)$$

The derivative of ΔG_w^* with respect to ε_{HB} may be obtained from the general expression (Sections 1.2 and 3.1)

$$\Delta G_w^* = - kT \ln \langle \exp(-\beta B_w) \rangle$$

$$= - kT \ln \left[\int d\mathbf{X}^N \exp(-\beta U_{N+1}) \bigg/ \int d\mathbf{X}^N \exp(-\beta U_N) \right]$$

$$(3.120)$$

Note that in equation (3.120) the range of integration over the locations of all particles is over the (TPN average) volume $\langle V \rangle$. See Section 3.1 for details.

We now use the general form for the total potential energy (3.104) of the systems of N and $N + 1$ particles to perform the differentiation of ΔG_w^* with respect to ε_{HB}. The result is

$$\frac{\partial \Delta G_w^*}{\partial \varepsilon_{HB}} = \frac{\int d\mathbf{X}^N \exp(-\beta U_{N+1}) \sum_{i<j}^{N+1} G(i,j)}{\int d\mathbf{X}^N \exp(-\beta U_{N+1})}$$

$$- \frac{\int d\mathbf{X}^N \exp(-\beta U_N) \sum_{j<j}^{N} G(i,j)}{\int d\mathbf{X}^N \exp(-\beta U_N)}$$

$$= \int d\mathbf{X}^N P(\mathbf{X}^N / \mathbf{X}_0) \sum_{i<j}^{N+1} G(i,j) - \int d\mathbf{X}^N P(\mathbf{X}^N) \sum_{i<j}^{N} G(i,j)$$

$$(3.121)$$

where the first integral is identified as a conditional average, i.e., an average over all the configurations of the N water molecules, given one water molecule at $\mathbf{X}_0$ [see also equation (3.114) of Section 3.8]. Here we use the shorthand notation $\Sigma_{i<j}$ for a sum over i and j such that $i < j$. Thus using equation (3.103), we rewrite (3.121) as

$$\frac{\partial \Delta G_w^*}{\partial \varepsilon_{HB}} = \langle HB \rangle_w^{N+1} - \langle HB \rangle_0^N \qquad (3.122)$$

Now we note that the conditional average number of HBs in a system of $N + 1$ particles, given that one of them is at a fixed configuration $\mathbf{X}_0$, will not be changed if we release the condition. In other words, the average number of HBs in the system of $N + 1$ water molecules is invariant to fixing the location and orientation of *one* of its molecules.

Hence we may write

$$\langle HB \rangle_w^{N+1} = \langle HB \rangle_0^{N+1} = \frac{N+1}{2} \langle \psi \rangle_0 \qquad (3.123)$$

$$\langle HB \rangle_0^N = \frac{N}{2} \langle \psi \rangle_0 \qquad (3.124)$$

So the derivative of ΔG_w^* may be simplified to obtain

$$\frac{\partial \Delta G_w^*}{\partial \varepsilon_{HB}} = \frac{N+1}{2} \langle \psi \rangle_0 - \frac{N}{2} \langle \psi \rangle_0 = \tfrac{1}{2} \langle \psi \rangle_0 \qquad (3.125)$$

and the approximate relation (3.119) is now rewritten as

$$\Delta G_{D_2O}^* - \Delta G_{H_2O}^* \cong \tfrac{1}{2} \langle \psi \rangle_0 (\varepsilon_D - \varepsilon_H) \qquad (3.126)$$

On the lhs we have a measurable quantity, the isotope effect on the solvation Gibbs energy of a water molecule in pure liquid water. On the rhs we have a measure of the structure of pure water $\langle \psi \rangle_0$ (see Section 3.7). This quantity may be evaluated if we can estimate the difference $\varepsilon_D - \varepsilon_H$. Some numerical examples of the application of equation (3.126) are presented in Section 2.11.

(2) *Dilute solution of inert solutes.* We now extend the treatment for pure liquid water to dilute aqueous solutions. We assume that the solute s does not form any HBs with the solvent molecules. Furthermore, we assume that the solute–solvent interaction is the same between S–H_2O and S–D_2O. The solvation Gibbs energy of s is

$$\Delta G_s^* = - kT \langle \exp(- \beta B_s) \rangle_0 \qquad (3.127)$$

where the average is over all the configurations of the solvent molecules, in the *TPN* ensemble.

Again viewing H_2O and D_2O as essentially the same liquid, except for a small difference in the HB energy parameter, we write the following expansion, similar to equation (3.119), namely

$$\Delta G_s^*(D_2O) = \Delta G_s^*(H_2O) + \frac{\partial \Delta G_s^*}{\partial \varepsilon_{HB}} (\varepsilon_D - \varepsilon_H) \qquad (3.128)$$

where

$$\frac{\partial \Delta G_s^*}{\partial \varepsilon_{HB}} = \frac{\int d\mathbf{X}^N \exp(-\beta U_N - \beta B_s) \sum_{i<j}^N G(i,j)}{\int d\mathbf{X}^N \exp(-\beta U_N - \beta B_s)}$$

$$- \frac{\int d\mathbf{X}^N \exp(-\beta U_N) \sum_{i<j}^N G(i,j)}{\int d\mathbf{X}^N \exp(-\beta U_N)}$$

$$= \langle HB \rangle_s^N - \langle HB \rangle_0^N \qquad (3.129)$$

Therefore, the expansion to first order in ε_{HB} becomes

$$\Delta G_s^*(D_2O) - \Delta G_s^*(H_2O) = (\langle HB \rangle_s^N - \langle HB \rangle_0^N)(\varepsilon_D - \varepsilon_H) \quad (3.130)$$

On the lhs of equation (3.130) we have a measurable quantity — the solvation Gibbs energy of a solute in H_2O and in D_2O. On the rhs we have the quantity $(\langle HB \rangle_s^N - \langle HB \rangle_0^N)$, which measures the change in the average number of HBs in a system of N water molecules, induced by the solvation process. This may be estimated, provided we have an estimated value for the difference $\varepsilon_D - \varepsilon_H$. Some numerical examples are presented in Section 2.10.

(3) *Dilute solutions of solutes that form HB with water.* For completeness, we discuss here the more complicated case of a solute that forms HB with the solvent.

We assume that the total potential energy of interaction of a system of N water molecules and one solute molecule s has the form

$$U(\mathbf{X}^N, \mathbf{X}_s) = U'(\mathbf{X}^N, \mathbf{X}_s) + \varepsilon_{HB} \sum_{i<j} G(i,j) + \varepsilon_{sw} \sum_{i=1}^N G_{sw}(s,i) \quad (3.131)$$

where the new feature is the HB part between the solute s and a water molecule w. As above, we assume here that the function G_{sw} is the same for H_2O and for D_2O. The HB energy parameters ε_{sH} and ε_{sD} for the pairs S–H_2O and S–D_2O will be distinguished. We have assumed for simplicity that the solute may form one kind of HB with a water molecule. In the most general case, a solute may form various kinds of HBs through different functional groups in s.

Using similar arguments as above, we expand ΔG_s^* to first order in both ε_{HB} and ε_{sw} to obtain

$$\Delta G_s^*(D_2O) = \Delta G_s^*(H_2O) + \frac{\partial \Delta G_s^*}{\partial \varepsilon_{HB}}(\varepsilon_D - \varepsilon_H) + \frac{\partial \Delta G_s^*}{\partial \varepsilon_{sw}}(\varepsilon_{sD} - \varepsilon_{sH}) \qquad (3.132)$$

Taking derivatives with respect to ε_{HB} and ε_{sw}, we obtain

$$\Delta G_s^*(D_2O) - \Delta G_s^*(H_2O) = (\langle HB \rangle_s^N - \langle HB \rangle_0^N)(\varepsilon_D - \varepsilon_H) + \langle \psi_s \rangle_s (\varepsilon_{sD} - \varepsilon_{sH})$$

$$(3.133)$$

We see that in this case we have an induced change in the average number of HBs in the system, as in equation (3.130), and an average number of HBs formed by the solute $\langle \psi_s \rangle_s$. Of course, evaluation of these quantities is more difficult since we need estimates of both $\varepsilon_D - \varepsilon_H$ and $\varepsilon_{sD} - \varepsilon_{sH}$.

We note two limiting cases of equation (3.133). If s does not form HB with a water molecule, then $\varepsilon_{sD} = \varepsilon_{sH} = 0$, and equation (3.133) reduces to (3.130). If, on the other hand, s itself is a water molecule, then $\varepsilon_{sD} = \varepsilon_D$ and $\varepsilon_{sH} = \varepsilon_H$, and the rhs of equation (3.133) reduces to [see also equations (3.123) and (3.124)]

$$(\langle HB \rangle_s^N - \langle HB \rangle_0^N + \langle \psi_s \rangle_s)(\varepsilon_D - \varepsilon_H)$$

$$= (\langle HB \rangle_s^{N+1} - \langle \psi \rangle_0 - \langle HB \rangle_0^N + \langle \psi \rangle_0)(\varepsilon_D - \varepsilon_H)$$

$$= \tfrac{1}{2}\langle \psi \rangle_0 (\varepsilon_D - \varepsilon_H) \qquad (3.134)$$

which is the same as equation (3.126).

3.10. SOLVATION GIBBS ENERGY AND KIRKWOOD–BUFF INTEGRALS

We derive here some useful relations between solvation Gibbs energies and integrals over the pair correlation functions. These integrals were introduced by Kirkwood and Buff,[6] and are sometimes referred to as Kirkwood–Buff integrals. For two species α and β the integral is defined as

$$G_{\alpha\beta} = \int_0^\infty [g_{\alpha\beta}(R) - 1]4\pi R^2 dR \qquad (3.135)$$

where $g_{\alpha\beta}$ is the pair correlation function for the species α and β. For spherical molecules $g_{\alpha\beta}$ is a function of R only. For nonspherical molecules the pair correlation function depends also on the relative orientation of the two particles. The function $g_{\alpha\beta}(R)$ used in the integrand is the average over all orientations of the pair of molcules; hence the result is a function of R only.[†]

We start with a pure one-component system, for which the well-known compressibility equation (see Chapter 3 of Ben-Naim[(4)]) is

$$\beta_T = \frac{1}{kT\rho} + \frac{G}{kT} \tag{3.136}$$

where β_T is the isothermal compressibility, ρ is the density, and G is an integral of the form (3.135), applied for a one-component system.

We use equation (3.136) together with the thermodynamic identity

$$\left(\frac{\partial\mu}{\partial\rho}\right)_T = (\beta_T\rho^2)^{-1} \tag{3.137}$$

to obtain

$$\left(\frac{\partial\mu}{\partial\rho}\right)_T = kT\left(\frac{1}{\rho} - \frac{G}{1+\rho G}\right) \tag{3.138}$$

Let ρ_0 be a very low density (later we let $\rho_0 \rightarrow 0$), so that at this density the system behaves as an ideal gas; therefore, the CP at ρ_0 is

$$\mu(\rho_0) = kT \ln \rho_0 \Lambda^3 \tag{3.139}$$

where, for simplicity, we assume that the molecules have no internal degrees of freedom; otherwise, a factor $-kT \ln q$ should be added to expression (3.139).

We now integrate equation (3.138) between ρ_0 and the final density ρ to obtain

$$
\begin{aligned}
\mu(\rho) &= \mu(\rho_0) + kT \int_{\rho_0}^{\rho} \left(\frac{1}{\rho'} - \frac{G}{1+\rho'G}\right) d\rho' \\
&= kT \ln \rho_0 \Lambda^3 + kT \ln(\rho/\rho_0) - kT \int_{\rho_0}^{\rho} \frac{G}{1+\rho'G} d\rho' \\
&= kT \ln \rho\Lambda^3 - kT \int_{0}^{\rho} \frac{G}{1+\rho'G} d\rho' \tag{3.140}
\end{aligned}
$$

where we put $\rho_0 = 0$ in the lower limit of the integral.

[†] Caution. We note that G in equation (3.135) is used for the Kirkwood–Buff integral and also for the Gibbs energy, such as in equation (3.141). However, the latter always appears in the form ΔG and therefore no confusion should arise.

From relation (3.140) we identify the solvation Gibbs energy of a molecule s in its pure liquid as

$$\Delta G_s^{*p} = - kT \int_0^{\rho_s} \frac{G_{ss}}{1 + \rho_s' G_{ss}} \, d\rho_s' \qquad (3.141)$$

where G_{ss}, the Kirkwood–Buff integral for the s–s pair, is, in general, a function of ρ_s'. Using the thermodynamical identity

$$\rho_s \beta_{T,s} = \left(\frac{\partial \rho_s}{\partial P} \right)_T \qquad (3.142)$$

together with the compressibility equation (3.136) (note that $\beta_{T,s}$ here is the same as β_T), we may rewrite equation (3.141) as

$$\Delta G_s^{*p} = - kT \int_0^{\rho_s} \frac{G_{ss}}{kT \rho_s' \beta_{T,s}} \, d\rho_s' = - \int_0^{\rho_s} G_{ss} \frac{\partial P}{\partial \rho_s'} \, d\rho_s'$$

$$= - \int_0^p G_{ss} dP' \qquad (3.143)$$

This is consistent with the equality (see also Section 1.5)

$$\Delta V_s^{*p} = \bar{V}_s - kT\beta_{T,s} = \frac{1 - kT\rho_s \beta_{T,s}}{\rho_s} = - G_{ss} \qquad (3.144)$$

where $\bar{V}_s$ is the molar (or molecular) volume of the pure liquid s. With this identification of $- G_{ss}$, the last equality on the rhs of equation (3.143) is simply the result of integrating the thermodynamic relationship

$$\frac{\partial \Delta G_s^{*p}}{\partial P} = \Delta V_s^{*p} = - G_{ss} \qquad (3.145)$$

The last equality on the rhs of equation (3.143) can be easily generalized to a dilute solution of s in a solvent, say w. In this case we have

$$\Delta V_s^{*0} = \bar{V}_s^0 - kT\beta_{T,w} \qquad (3.146)$$

where $\bar{V}_s^0$ is the partial molecular volume of s in the limit of a very dilute solution in w, and $\beta_{T,w}$ is the isothermal compressibility of the pure solvent, w.

We now use the expression for $\bar{V}_s^0$ from the Kirkwood–Buff theory (see, for instance, Chapter 4 in Ben-Naim[4]):

$$\bar{V}_s^0 = \frac{1 + \rho_w^0(G_{ww}^0 - G_{sw}^0)}{\rho_w^0} \tag{3.147}$$

which, together with the compressibility equation for the solvent and with equation (3.146), yields the result

$$\Delta V_s^{*0} = - G_{sw}^0 \tag{3.148}$$

or

$$\Delta G_s^{*0} = \int_0^p \Delta V_s^{*0} dP = - \int_0^p G_{sw}^0 dP \tag{3.149}$$

which is the analogue of equation (3.143) for the solvation Gibbs energy of s in the limit of a very dilute solution in w.

Another relation that has been derived,[55] the analogue of equation (3.141), is

$$\Delta G_s^{*0} = - kT \int_0^{\rho_w} \frac{G_{sw}^0}{1 + \rho_w' G_{sw}^0} \, d\rho_w' \tag{3.150}$$

The last equation can also be derived from equation (3.149) by changing variables from P to ρ_w and using relation (3.142). We note that ΔG_s^{*p} and ΔG_s^{*0} tend to zero when $\rho_s \to 0$, in equation (3.141) or $\rho_w \to 0$ in equation (3.150), a feature that was discussed in Section 1.4.

3.11. PREFERENTIAL SOLVATION IN A TWO-COMPONENT SOLVENT†

Consider a simple solute s in a very dilute solution in a solvent. The solvent consists of two components A and B, e.g., argon in water ethanol mixtures. In this section we address ourselves to the following question: given a solvent mixture of bulk composition $x_A = N_A/(N_A + N_B)$, what can we say about the local composition of the solvent in the immediate

---

† In this section we discuss the subject of preferential solvation (PS) in a three-component system, a solute and a two-component solvent. However, we note that the concept of PS can also be used in a two-component system of A and B. At any composition, one can ask about the PS of either A or B with respect to the two components in their surroundings. The answer to this question can be given within the concept of solvation as defined in this book, but not in any of the conventional approaches to solvation thermodynamics. This topic will be discussed in a future article.

vicinity of the solute? In other words, suppose we choose any element of volume dV in the liquid, the average numbers of A and B molecules in dV being $\rho_A dV$ and $\rho_B dV$, respectively. Now, we consider an element of volume dV at a distance R from the center of s. What are the average numbers of A and B molecules in this element of volume? The answer is simple:

$$\langle dN_A \rangle = \rho_A g_{As}(R) dV \tag{3.151}$$

$$\langle dN_B \rangle = \rho_B g_{Bs}(R) dV \tag{3.152}$$

where g_{As} and g_{Bs} are the pair correlation functions for the A, s and B, s pairs, respectively. Here, the pair correlation function is presumed to be a function of R only. If s is a more complex molecule, say a protein, then we can ask a more detailed question regarding the average numbers of A and B molecules as some distance R and at a specific direction relative to the orientation of s. Some aspect of this problem will be discussed in Section 3.19.

At present, there is neither a theoretical nor an experimental means of computing $g_{As}(R)$ and $g_{Bs}(R)$ at any distance R of the order of magnitude of a molecular diameter of s. Therefore, we turn to a more modest question. Instead of looking at any specific distance R, we shall examine the relative *average* excess, or deficiency, of A and B in the entire surroundings of s. The relevant quantity is the Kirkwood–Buff integral, which has been introduced in Section 3.10:

$$G_{\alpha\beta} = \int_0^\infty [g_{\alpha\beta}(R) -]4\pi R^2 dR \tag{3.153}$$

The significance of this quantity with respect to the question of preferential solvation is the following. The quantity $\rho_A g_{As}(R)4\pi R^2 dR$ is the average number of A particles in the element of volume $4\pi R^2 dR$ at distance R from the center of s. On the other hand, $\rho_A 4\pi R^2 dR$ is the average number of A molecules in the same element of volume, but taken relative to an arbitrary center in the liquid. Therefore, $\rho_A[g_{As}(R) - 1]4\pi R^2 dR$ measures the excess, or deficiency, of A molecules in the spherical shell of volume $4\pi R^2 dR$ around s relative to the same spherical shell, but at an arbitrary location in the liquid. The quantity $\rho_A G_{As}$, according to the definition (3.153), is simply the overall excess or deficiency of A molecules in the entire volume around s [since the function $g_{As}(R)$ normally decays to unity at distances of a few molecular diameters, the integral (3.153) gets its major contribution from a relatively short distance around s].

Thus instead of examining the quantities defined in expressions

(3.151) and (3.152), we shall examine the less-detailed quantities:

$$\delta N_{A,s} = \rho_A G_{As} \tag{3.154}$$

and

$$\delta N_{B,s} = \rho_B G_{Bs} \tag{3.155}$$

Since ρ_A and ρ_B are presumed to be given for any given composition of the solvent mixture, the main problem is to find a way of computing G_{As} and G_{Bs} from experimental data. The following procedure suggests a route by which we can compute the required quantities. In the following we shall restrict ourselves to very dilute solutions of s in a mixture of A and B of any composition x_A.

We denote by l the liquid mixture of A and B; the CP of s in l is given by

$$\mu_s^l = \mu_s^{*l} + kT \ln \rho_s^l \Lambda_s^3 \tag{3.156}$$

The derivative of μ_s^l with respect to, say, N_A is

$$\left(\frac{\partial \mu_s^l}{\partial N_A}\right)_{T,P,N_B,N_s} = \left(\frac{\partial \mu_s^{*l}}{\partial N_A}\right)_{T,P,N_B,N_s} - \frac{kT}{V}\,\bar{V}_A \tag{3.157}$$

where $\bar{V}_A$ is the partial molar volume of A in the mixture. The derivative in equation (3.157) may be expressed in terms of Kirkwood–Buff integrals. The following is an exact result:[56]

$$\left(\frac{\partial \mu_s^l}{\partial N_A}\right)_{T,P,N_B,N_s} = \frac{kT}{\rho_s \rho_A V}\,\frac{|E(\text{s},A)|}{|D|} \tag{3.158}$$

where $|E(\text{s},A)|$ and $|D|$ are two determinants, the definitions of which are available[56] for any mixture. Here we shall be interested only in the case of very dilute solutions of s in the mixture. For this case we have ($\rho_s^l \to 0$)

$$
|E(\text{s},A)| = \begin{vmatrix} 1 & 0 & 1 \\ G_{sA} & 1 & G_{AB} \\ G_{sB} & 1 & G_{BB} + \rho_B^{-1} \end{vmatrix}
$$

$$
= \frac{1}{\rho_B}(1 + \rho_B G_{BB} - \rho_B G_{AB}) + G_{sA} - G_{sB}
$$

$$
= \eta \bar{V}_A \rho_B^{-1} + G_{sA} - G_{sB} \tag{3.159}
$$

where η is defined (for the mixture of A and B) by

$$\eta = \rho_A + \rho_B + \rho_A\rho_B(G_{AA} + G_{BB} - 2G_{AB}) \qquad (3.160)$$

and

$$|D| = \begin{vmatrix} 0 & 1 & 1 & 1 \\ 1 & G_{ss} + \rho_s^{-1} & G_{sA} & G_{sB} \\ 1 & G_{sA} & G_{AA} + \rho_A^{-1} & G_{AB} \\ 1 & G_{sB} & G_{AB} & G_{BB} + \rho_B^{-1} \end{vmatrix} \qquad (3.161)$$

$$\xrightarrow[(\rho_s \to 0)]{} \frac{-\eta}{\rho_s\rho_A\rho_B} \qquad (3.162)$$

Substitution of expressions (3.159) and (3.162) in equation (3.158) yields

$$\left(\frac{\partial \mu_s^l}{\partial N_A}\right)_{T,P,N_B,N_s} = \frac{-kT}{V\eta} [\eta\bar{V}_A + \rho_B(G_{sA} - G_{sB})] \qquad (3.163)$$

From equations (3.157) and (3.163) we obtain

$$\left(\frac{\partial \mu_s^{*l}}{\partial N_A}\right)_{T,P,N_B,N_s} = \frac{-kT}{V} \frac{\eta\bar{V}_A + \rho_B(G_{sA} - G_{sB})}{\eta} + \frac{kT}{V}\bar{V}_A \qquad (3.164)$$

However, for practical purposes it is more convenient to transform into a derivative with respect to the mole fraction x_A:

$$\frac{\partial \mu_s^{*l}}{\partial N_A} = \frac{\partial \mu_s^{*l}}{\partial x_A} \frac{\partial x_A}{\partial N_A} = \frac{\partial \mu_s^{*l}}{\partial x_A} \frac{N_B}{(N_A + N_B)^2} \qquad (3.165)$$

and hence the final result is obtained in the form

$$\lim_{\rho_s \to 0} \left(\frac{\partial \mu_s^{*l}}{\partial x_A}\right)_{T,P} = \frac{kT(\rho_A + \rho_B)^2}{\eta} (G_{Bs} - G_{As}) \qquad (3.166)$$

Since μ_s^{*g} in an ideal gas is independent of x_A, we could also write relation (3.166) as

$$\lim_{\rho_s \to 0} \left(\frac{\partial \Delta G_s^{*l}}{\partial x_A}\right)_{P,T} = \frac{kT(\rho_A + \rho_B)^2}{\eta} (G_{Bs} - G_{As}) \qquad (3.167)$$

Thus by measuring the solvation Gibbs energy of s in a series of mixtures of A and B, one can determine the difference $G_{Bs} - G_{As}$ from equation (3.167). (The quantity η may be determined independently from the properties of the two-component system of A and B.[57,58]) However, it will be seen below that both G_{Bs} and G_{As} can actually be computed by adding one more equation to (3.167). Before deriving the second equation, we note that relation (3.167) is useful in the case of a solute for which the solvation Gibbs energy may be determined. If we are interested in a protein solute, then equation (3.167) is impractical. However, we can still measure the relative solvation Gibbs energy of s in l relative to, say, pure A, i.e.,

$$\Delta\Delta G_s^* \equiv \Delta G_s^{*l} - \Delta G_s^{*A} = kT \ln(\rho_s^A/\rho_s^l)_{eq} \tag{3.168}$$

Hence by measuring the density of s in l and in pure A at equilibrium with respect to the pure, say solid, s, we can determine $\Delta\Delta G_s^*$ from relation (3.168). But since ΔG_s^{*A} is independent of x_A, we can apply relation (3.168) in equation (3.167) to obtain

$$\lim_{\rho_s \to 0} \left[\frac{\partial(\Delta\Delta G_s^*)}{\partial x_A} \right] = \frac{kT(\rho_A + \rho_B)^2}{\eta} (G_{Bs} - G_{As}) \tag{3.169}$$

which is the useful relation for complex solutes, such as proteins. (We note that when $G_{Bs} = G_{As}$, the solvation Gibbs energy becomes independent of the composition of the solvent.)

The second equation that we shall now derive is for the partial molar volume of s at infinite dilution in the solvent l, of composition x_A.

From the Kirkwood–Buff theory (for details see Ben-Naim[4]) we have the following equation:

$$\mu_{\alpha\beta} = \left(\frac{\partial\mu_\alpha}{\partial N_\beta} \right)_{P,T,N'} = \frac{kTB^{\alpha\beta}}{V|B|} - \frac{\bar{V}_\alpha \bar{V}_\beta}{\beta_T V} \tag{3.170}$$

where $|B|$ is the determinant defined by

$$|B| = \begin{vmatrix} \rho_s + \rho_s^2 G_{ss} & \rho_s\rho_A G_{As} & \rho_s\rho_B G_{Bs} \\ \rho_s\rho_A G_{As} & \rho_A + \rho_A^2 G_{AA} & \rho_A\rho_B G_{AB} \\ \rho_s\rho_B G_{Bs} & \rho_A\rho_B G_{AB} & \rho_B + \rho_B^2 G_{BB} \end{vmatrix} \tag{3.171}$$

which has the following limiting behavior at $\rho_s \to 0$:

$$|B| \to \rho_s\rho_A\rho_B\xi \tag{3.172}$$

with ξ defined by

$$\xi = 1 + \rho_A G_{AA} + \rho_B G_{BB} + \rho_A \rho_B (G_{AA} G_{BB} - G_{AB}^2) \qquad (3.173)$$

The quantities $B^{\alpha\beta}$ are the cofactors of the determinant $|B|$, which for the limit $\rho_s \to 0$ are given by

$$B^{sA} \to -\rho_s \rho_A \rho_B [G_{sA}(1 + \rho_B G_{BB}) - G_{sB} \rho_B G_{AB}] \qquad (3.174)$$

$$B^{sB} \to -\rho_s \rho_A \rho_B [G_{sB}(1 + \rho_A G_{AA}) - G_{sA} \rho_A G_{AB}] \qquad (3.175)$$

$$B^{ss} \to \rho_A \rho_B \xi \qquad (3.176)$$

Using the Gibbs–Duhem equation

$$\rho_A \mu_{sA} + \rho_B \mu_{sB} + \rho_s \mu_{ss} = 0 \qquad (3.177)$$

and the Kirkwood–Buff results for the isothermal compressibility and the partial molar volumes of A and B, namely

$$\beta_T = \xi / kT\eta \qquad (3.178)$$

$$\bar{V}_A = [1 + \rho_B (G_{BB} - G_{AB})]/\eta \qquad (3.179)$$

$$\bar{V}_B = [1 + \rho_A (G_{AA} - G_{AB})]/\eta \qquad (3.180)$$

we obtain the following equality:

$$\rho_A \left(\frac{kTB^{sA}}{V|B|} - \frac{\bar{V}_s \bar{V}_A}{\beta_T V} \right) + \rho_B \left(\frac{kTB^{sB}}{V|B|} - \frac{\bar{V}_s \bar{V}_B}{\beta_T V} \right) + \rho_s \left(\frac{kTB^{ss}}{V|B|} - \frac{\bar{V}_s^2}{\beta_T V} \right) = 0$$
$$(3.181)$$

which may be reduced to

$$\xi - \rho_A G_{sA} \bar{V}_A \eta - \rho_B G_{sB} \bar{V}_B \eta - \rho_A \bar{V}_s \bar{V}_A \eta - \rho_B \bar{V}_s \bar{V}_B \eta = 0 \qquad (3.182)$$

We may eliminate $\bar{V}_s$ (for $\rho_s \to 0$) from the latter equation to obtain the final result:

$$\bar{V}_s^0 = \lim_{\rho_s \to 0} \bar{V}_s = kT\beta_T - \rho_A \bar{V}_A G_{sA} - \rho_B \bar{V}_B G_{sB} \qquad (3.183)$$

In equation (3.167) [or (3.169)] and equation (3.183) all the quantities β_T, η, ρ_A, ρ_B, $\bar{V}_A$, $\bar{V}_B$, $\bar{V}_s^0$, and $\partial \Delta G_s^* / \partial X_A$ are experimentally determinable. Hence these two equations may be used to eliminate the required quantities G_{sB} and G_{sA}. Thus by denoting

$$a = \frac{\partial \Delta G_s^{*l}}{\partial X_A}, \quad b = \frac{kT(\rho_A + \rho_B)^2}{\eta}, \quad \text{and} \quad c = kT\beta_T \quad (3.184)$$

we may solve for G_{As} and G_{Bs} in the form

$$G_{sB} = c - \bar{V}_s^0 + \frac{a}{b}\,\rho_A \bar{V}_A \quad (3.185)$$

$$G_{sA} = c - \bar{V}_s^0 - \frac{a}{b}\,\rho_B \bar{V}_B \quad (3.186)$$

which are the required quantities.

3.12. SOLVATION THERMODYNAMICS OF COMPLETELY DISSOCIABLE SOLUTES

In this section we develop the statistical mechanical relation between the solvation thermodynamics of a pair of ions and experimental quantities. This relation is the basic theoretical background for the calculations reported in Section 2.17.

We consider a molecule which is in a state of a dimer D, in the gaseous phase, and when it is introduced into the liquid it completely dissociates into two fragments A and B.[†] Of foremost importance is the case of ionic solutes, e.g., D may be HCl, then A and B are H^+ and Cl^-, respectively. For simplicity we assume that A and B do not have any internal degrees of freedom. The generalization to the case of multi-ionic solutes and polynuclear ions is quite straightforward. These cases will be discussed briefly at the end of this section.

The relevant solvation process has been defined in Section 1.6. Since the particles are presumed not to possess any internal degrees of freedom, the Gibbs energy of solvation of the pair of groups (or ions) is simply the coupling work of A and B to the liquid phase l; thus

$$\Delta G_{AB}^* = W(A, B \mid l) \quad (3.187)$$

Also, for simplicity we shall use below the T, V, N ensemble. Hence we shall derive an expression for ΔA_{AB}^*. However, using essentially the

---

[†] The assumption that D is in dimeric form in the gaseous state in not essential. However, this particular case was treated in Section 2.17, for which the relevant data are available.

same argument as given in Section 3.1, we can easily show that ΔA_{AB}^* at T, V constant is the same quantity as ΔG_{AB}^* at T, P, constant.

Consider now a system of N_w solvent molecules and N_D solute molecules contained in a volume V and at temperature T. By "solute" we mean those molecules D for which we wish to evaluate their solvation thermodynamic quantities. The "solvent," which is usually water, may be any liquid or any mixture of liquids, and could contain any number of other solutes besides D. In the most general case, N_w will be the total number of molecules in the system except those that are counted as "solutes" in N_D. However, for notational simplicity we shall treat only two-component systems, w and D.

The procedure for obtaining the solvation Helmholtz energy ΔA_{AB}^* is essentially that described in a qualitative manner in Section 1.6, using the diagram of Figure 1.3.

We shall need the following partition functions. First, we need the partition function of the system of N_w solvent molecules and N_D solutes, at given V, T without any further restrictions:

$$Q(T, V, N_w, N_D)$$

$$= \frac{q_w^{N_w} q_A^{N_A} q_B^{N_B} \int d\mathbf{X}^{N_w} d\mathbf{X}^{N_A} d\mathbf{X}^{N_B} \exp[-\beta U(N_w, N_A, N_B)]}{\Lambda_w^{3N_w} \Lambda_A^{3N_A} \Lambda_B^{3N_B} N_A! \, N_B! \, N_w!} \tag{3.188}$$

where we have assumed that the solute molecules are completely dissociated into A and B in the liquid; $U(N_w, N_A, N_B)$ is the total potential energy of interaction among the N_w, N_A, and N_B molecules at a specific configuration (with $N_D = N_A = N_B$). If the solvent contains more than one component, then q_w and Λ_w should be interpreted as products of the corresponding quantities for all the components present in the solvent.

We next add one dimer D, or equivalently one A and one B, to the liquid in two different ways: first, without any further restrictions, and second, with the constraint that A and B be placed at two fixed positions $\mathbf{R}_A$ and $\mathbf{R}_B$, respectively. The corresponding partition functions are

$$Q(T, V, N_w, N_D + 1)$$

$$= \frac{q_w^{N_w} q_A^{N_A+1} q_B^{N_B+1} \int d\mathbf{R}_A d\mathbf{R}_B \int d\mathbf{X}^{N_w+N_A+N_B} \exp[-\beta U(N_w, N_D+1)]}{\Lambda_w^{3N_w} \Lambda_A^{3(N_A+1)} \Lambda_B^{3(N_B+1)} N_w! \, (N_A+1)! \, (N_B+1)!} \tag{3.189}$$

and

$$Q(T, V, N_w, N_D + 1; \mathbf{R}_A, \mathbf{R}_B)$$

$$= \frac{q_w^{N_w} q_A^{N_A + 1} q_B^{N_B + 1} \int d\mathbf{X}^{N_w + N_A + N_B} \exp[-\beta U(N_w, N_D + 1)]}{\Lambda_w^{3N_w} \Lambda_A^{3N_A} \Lambda_B^{3N_B} N_w! \, N_A! \, N_B!} \tag{3.190}$$

Note the difference in the various factors in equations (3.189) and (3.190). In equation (3.189) we have two more integrations over $\mathbf{R}_A$ and $\mathbf{R}_B$, and also one more Λ_A^3 and Λ_B^3. In addition, we have $(N_A + 1)!$ and $(N_B + 1)!$ in equation (3.189), but only $N_A!$ and $N_B!$ in (3.190). All these differences arise from the constraint we have imposed on the locations $\mathbf{R}_A$ and $\mathbf{R}_B$ in equation (3.190). The total potential energies of interaction in the two systems are related by the equation

$$U(N_w, N_D + 1) = U(N_w, N_D) + U_{AB}(\mathbf{R}_A, \mathbf{R}_B) + B_D(\mathbf{R}_A, \mathbf{R}_B) \tag{3.191}$$

where the configuration of the system, except for $\mathbf{R}_A$ and $\mathbf{R}_B$, is denoted symbolically by (N_w, N_D) and $(N_w, N_D + 1)$. The quantity $U_{AB}(\mathbf{R}_A, \mathbf{R}_B)$ is the direct interaction potential between A and B, being at $\mathbf{R}_A$ and $\mathbf{R}_B$, respectively; $B_D(\mathbf{R}_A, \mathbf{R}_B)$ is the "binding energy," i.e., the total interaction energy between the pair A and B at $(\mathbf{R}_A, \mathbf{R}_B)$ and all the other particles in the system being at some specific configuration (the latter is omitted from the notation for simplicity).

We are now ready to derive the expression for the CP of the solute D in the liquid l. This is obtained from equations (1.188) and (1.189), i.e.,

$$\mu_D^l = -kT \ln[Q(T, V, N_w, N_D + 1)/Q(T, V, N_w, N_D)]$$

$$= -kT \ln \left\{ \left[q_A q_B \int d\mathbf{R}_A d\mathbf{R}_B \exp[-\beta U_{AB}(\mathbf{R}_A, \mathbf{R}_B)] \right. \right.$$

$$\left. \times \int d\mathbf{X}^{N_w + N_A + N_B} \exp[-\beta(N_w, N_D) - \beta B_D(\mathbf{R}_A, \mathbf{R}_B)] \right]$$

$$\left. \times \left[\Lambda_A^3 \Lambda_B^3 (N_A + 1)(N_B + 1) \int d\mathbf{X}^{N_w + N_A + N_B} \exp[-\beta U(N_w, N_D)] \right]^{-1} \right\}$$

$$= -kT \ln \left\{ \frac{q_A q_B}{\Lambda_A^3 \Lambda_B^3 (N_A + 1)(N_B + 1)} \right.$$

$$\left. \times \int d\mathbf{R}_A d\mathbf{R}_B \exp[-\beta U_{AB}(\mathbf{R}_A, \mathbf{R}_B)] \langle \exp(-\beta B_D) \rangle_* \right\} \tag{3.192}$$

where the symbol $\langle \ \rangle_*$ indicates an average over all configurations of the $N_w + N_A + N_B$ particles, excluding only the solvaton AB at $\mathbf{R}_A$ and $\mathbf{R}_B$.

Equation (3.192) may be transformed into a simpler and more convenient form as follows. The pair distribution function for the species A and B in the liquid is defined by[49]

$$\rho_{AB}^{(2)}(\mathbf{R}_A, \mathbf{R}_B)$$

$$= (N_A + 1)(N_B + 1) \frac{\int d\mathbf{X}^{N_w + N_A + N_B} \exp[-\beta U(N_w, N_D + 1)]}{\int d\mathbf{X}^{N_w + (N_A + 1) + (N_B + 1)} \exp[-\beta U(N_w, N_D + 1)]}$$

$$(3.193)$$

If the pair correlation function at infinite separation is denoted by $\rho_{AB}^{(2)}(\infty)$, then we can write the following ratio:

$$\frac{\rho_{AB}^{(2)}(\mathbf{R}_A, \mathbf{R}_B)}{\rho_{AB}^{(2)}(\infty)}$$

$$= \frac{\int d\mathbf{X}^{N_w + N_A + N_B} \exp[-\beta U(N_w, N_D) - \beta B_D(\mathbf{R}_A, \mathbf{R}_B) - \beta U_{AB}(\mathbf{R}_A, \mathbf{R}_B)]}{\int d\mathbf{X}^{N_w + N_A + N_B} \exp[-\beta U(N_w, N_D) - \beta B_D(\infty) - \beta U_{AB}(\infty)]}$$

$$(3.194)$$

where we have used equation (3.191) and also introduced the notations $B_D(\infty)$ and $U_{AB}(\infty)$ for the binding energy and interaction energy at infinite separation. Using now the same probability distribution as used in relation (3.192), we may rewrite equation (3.194) in the form

$$\frac{\rho_{AB}^{(2)}(\mathbf{R}_A, \mathbf{R}_B)}{\rho_{AB}^{(2)}(\infty)} = \frac{\exp[-\beta U_{AB}(\mathbf{R}_A, \mathbf{R}_B)]\langle \exp[-\beta B_D(\mathbf{R}_A, \mathbf{R}_D)])\rangle_*}{\exp[-\beta U_{AB}(\infty)]\langle \exp[-\beta B_D(\infty)]\rangle_*} \quad (3.195)$$

Equation (3.195) is now introduced into relation (3.192) to yield

$$\mu_D^l = \tag{3.196}$$

$$-kT \ln \left\{ \frac{q_A q_B \int d\mathbf{R}_A d\mathbf{R}_B \rho_{AB}^{(2)}(\mathbf{R}_A \mathbf{R}_B) \exp[-\beta U_{AB}(\infty)]\langle \exp[-\beta B_D(\infty)]\rangle_*}{\Lambda_A^3 \Lambda_B^3 (N_A + 1)(N_B + 1)\rho_{AB}^{(2)}(\infty)} \right\}$$

Since we define our zero potential energy at $R_{AB} = \infty$, we have $U_{AB}(\infty) = 0$. Also, we have $\rho_{AB}^{(2)}(\infty) = \rho_A \rho_B$. The normalization condition[4,49]

$$\int d\mathbf{R}_A d\mathbf{R}_B \rho_{AB}^{(2)}(\mathbf{R}_A, \mathbf{R}_B) = (N_A + 1)(N_B + 1) \tag{3.197}$$

in now employed to obtain the final form:

$$\mu_D^l = - kT \ln\langle\exp[-\beta B_D(\infty)]\rangle_* + kT \ln(\rho_A\rho_B\Lambda_A^3\Lambda_B^3/q_Aq_B) \qquad (3.198)$$

Note that in relation (3.192) we wrote μ_D^l as an average over all possible positions of A and B; the simplification achieved in equation (3.198) was rendered possible because there is equilibrium among all pairs of A and B at any specific configuration $\mathbf{R}_A$, $\mathbf{R}_B$. This fact allows us to choose one configuration $R_{AB} = \infty$ and use it to obtain the simplified form of the CP, μ_D^l.

We can now also identify the liberation Helmholtz energy of the pair A and B. For any specific configuration this may be obtained from the ratio of the partition functions (3.189) and (3.190), i.e.,

$$\Delta A(\text{Lib}) = - kT \ln[Q(T, V, N_w, N_D + 1)/Q(T, V, N_w, N_D + 1; \mathbf{R}_A, \mathbf{R}_B)]$$
$$= - kT$$

$$\times \ln\left\{\frac{\int d\mathbf{R}_A d\mathbf{R}_B \int d\mathbf{X}^{N_w+N_A+N_B}\exp[-\beta U(N_w, N_D + 1)]}{\Lambda_A^3\Lambda_B^3(N_A + 1)(N_B + 1)\int d\mathbf{X}^{N_w+N_A+N_B}\exp[-\beta U(N_w, N_D + 1)]}\right\}$$

$$= kT \ln[\Lambda_A^3\Lambda_B^3\rho^{(2)}(\mathbf{R}_A, \mathbf{R}_B)] \qquad (3.199)$$

which is a generalization of the expression for the liberation Helmholtz energy of one particle (Section 3.1). Here, the liberation Helmholtz energy depends on the particular configuration of A and B from which these particles are being released. For our application we need only the liberation energy at infinite separation, namely

$$\Delta A(\text{Lib}, \infty) = kT \ln(\Lambda_A^3\Lambda_B^3\rho_A\rho_B) \qquad (3.200)$$

Using relation (3.200) in equation (3.198), we may identify the solvation Helmholtz energy of the pair A and B, as defined in Section 1.6, i.e.,

$$\Delta A_{AB}^* = - kT \ln\langle\exp[-\beta B_D(\infty)]\rangle_* - kT \ln q_A q_B \qquad (3.201)$$

In all the examples given in Section 2.17, we assume also that the ions are simple, in the sense that $q_A = q_B = 1$.

In order to obtain a relation between the solvation Helmholtz energy and experimentally measurable quantities, we need to follow the various steps along the diagram in Figure 1.3. To do so we need the expression for the CP of D in the gaseous phase (where it is assumed that it exists only in dimeric form). This is simply

$$\mu_D^g = kT \ln \rho_D^g \Lambda_D^3 q_D^{-1} \tag{3.202}$$

where ρ_D^g is the number density and Λ_D^3 the momentum partition function of D; q_D includes the rotational, vibrational, and electronic partition functions of D. For all practical purposes, we may assume that the dimer D at room temperature is in its electronic and vibrational ground states. We also assume that the rotation may be treated classically; hence we write

$$q_D = q_{rot} \exp(-\beta \tfrac{1}{2}h\nu) \exp(-\beta \varepsilon_{el})$$

$$= q_{rot} \exp(\beta D_0) \tag{3.203}$$

where $\varepsilon_{el} = U^{AB}(\sigma) - U_{AB}(\infty)$ is the energy required to bring A and B from infinite separation to the equilibrium distance $R_{AB} = \sigma$. The experimental dissociation energy, as measured relative to the vibrational zero-point energy, is

$$D_0 = -\tfrac{1}{2}h\nu - \varepsilon_{el} = -\tfrac{1}{2}h\nu - [U_{AB}(\sigma) - U_{AB}(\infty)] \tag{3.204}$$

where $\tfrac{1}{2}h\nu$ is the zero-point energy for the vibration of D. We note that in the case of ionic solution we can either separate the two *ions* from σ to ∞, or first separate the two neutral *atoms* and then add the ionization energy and electron affinity of the cation and anion, respectively. The latter procedure is adopted in the numerical calculations reported in Section 2.17.[46]

We now have all the required quantities to follow the stepwise process as indicated in the diagram of Figure 1.3. Hence using the equilibrium condition for D in the two phases, and putting $q_A = q_B = 1$, we have

$$0 = \mu_D^l - \mu_D^g = [-kT \ln \rho_D^g \Lambda_D^3 q_{rot}^{-1} - \tfrac{1}{2}h\nu] + [U_{AB}(\infty) - U_{AB}(\sigma)]$$

$$- kT \ln \langle \exp[-\beta B_D(\infty)]\rangle_* + [kT \ln \rho_A \rho_B \Lambda_A^3 \Lambda_B^3] \tag{3.205}$$

In relation (3.205), as in Section 1.6 where we discussed this equation without proof, each term in squared brackets corresponds to the Helmholtz energy change for each step along the thick arrows in the diagram of Figure 1.3. From equation (3.205) we can eliminate the required Helmholtz energy of solvation of the pair AB to obtain the final relation that has been used in the computations reported in Section 2.17, namely

$$\Delta A_{AB}^* = - kT \ln \langle \exp[-\beta B_D(\infty)] \rangle_*$$

$$= -D_0 + kT \ln \left[\left(\frac{\Lambda_D^3}{\Lambda_A^3 \Lambda_B^3 q_{rot}} \right) \left(\frac{\rho_D^g}{\rho_A^l \rho_B^l} \right)_{eq} \right] \quad (3.206)$$

We see that in this case we need not only the number densities of D and of A and B at equilibrium, but also the molecular quantities Λ_A^3, Λ_B^3, Λ_D^3, q_{rot}, and D_0. We now briefly discuss the more general case of solvation of a multi-ionic solute.

Consider a solute s, which is presumed to be completely dissociated into n_s simple particles (or ions) in the liquid phase. We denote the composition of the solute s by the vector $\mathbf{n}_s = (n_A, n_B, n_C, \ldots)$ such that n_α is the number of α particles contained in s. Thus for $SnCl_3Br$ we have $n_{Sn} = 1$, $n_{Cl} = 3$, $n_{Br} = 1$. The total number of particles contained in s is $n_s = \Sigma_{\alpha=1}^r n_\alpha$, where r is the number of different species. We need not indicate the electrical charge on each particle. Of course, if s dissociates into ions, the condition of electrical neutrality of the total n_s particles will be required.

The CP of s in an ideal-gas phase, where it is presumed to exist as a single entity,[†] is

$$\mu_s^{ig} = kT \ln \rho_s^g \Lambda_s^3 q_s^{-1} \quad (3.207)$$

where q_s includes all the internal partition functions of an s molecule.

The CP of s in the liquid is more complicated. Following a similar procedure to that described above for a dimeric solute, we write for the more general case (more details are given elsewhere[(46)])

$$\mu_s^l = -kT \ln \langle \exp[-\beta B_s(\infty)] \rangle_* + kT \ln \prod_{\alpha=1}^r \rho_\alpha^{n_\alpha} \Lambda_\alpha^{3n_\alpha} q_\alpha^{-n_\alpha} \quad (3.208)$$

where $B_s(\infty)$ is the binding energy of all the n_s particles at fixed positions and infinite separation from each other. The significance of the average $\langle \ \rangle_*$ is the same as in equation (3.192).

As in the case of dimeric solutes, in order to measure the solvation Helmholtz energy we need to follow a diagram similar to that described in Section 1.3. Imposing the equilibrium condition and using equations

---

[†] As in the case of 1:1 electrolytes, this assumption is not essential.

(3.207) and (3.208), we obtain

$$0 = \mu_s^l - \mu_s^g$$

$$= \left(-kT \ln \rho_s^g \Lambda_s^g q_s^{-1} \Big/ \prod_{\alpha=1}^{r} q_\alpha^{-n_\alpha} \right)$$

$$+ (-kT \ln \langle \exp[-\beta B_s(\infty)] \rangle_*) + \left(kT \ln \prod_{\alpha=1}^{r} \rho_\alpha^{n_\alpha} \Lambda_\alpha^{3n_\alpha} \right) \qquad (3.209)$$

The solvation Helmholtz energy of the solute s can be eliminated from this equation. It is clear that, besides the equilibrium densities, we need a large number of molecular quantities in order to calculate the solvation Helmholtz energy of s.

We also note that if the ions are not simple (such as SO_4^{-2} and PO_4^{-3}) we must include in q_α all the relevant internal degrees of freedom of the α species. Also, if some of the ions have internal rotations (say in $CH_3CH_2CH_2COONa$), then the solvation Helmholtz energy obtained by the method used above should be interpreted as an average over all possible conformations of the relevant ions. For more details on this conformational average, the reader is referred to Section 3.2.

3.13. SINGLE ION CONTRIBUTIONS TO THE SOLVATION GIBBS ENERGY OF AN IONIC SOLUTE

In the conventional theory of ionic solvation it has been assumed customarily that the (conventional) solvation Gibbs energy may be written as a sum over single ion contributions. For example, the standard Gibbs energy of solution of KCl has been written as

$$\Delta G_{KCl}^\circ = \Delta G_{K^+}^\circ + \Delta G_{Cl^-}^\circ \qquad (3.210)$$

Thermodynamics alone does not provide any supporting argument to a splitting of the type (3.210), nor does it offer any method of evaluating the single ion quantities $\Delta G_{K^+}^\circ$ and $\Delta G_{Cl^-}^\circ$. The qualitative reasoning underlying the split in equation (3.210) is the following. Since we are dealing with very dilute solutions, the average ion–ion distance is very large; therefore, it is likely that each solute is solvated independently, and therefore the solvation Gibbs energy of KCl will be the sum of the two contributions due to K^+ and to Cl^{-1}. Clearly, this argument will not hold, from the thermodynamical point of view, for high concentrations of the ionic solute. In the latter the average ion–ion distance might be small enough to lead to ion–ion correlation, and hence dependence.

However, the following statistical mechanical argument is quite different, and holds for any concentration of the solute. We note that in the definition of the solvation Gibbs energy of the dimer (as well as in the more general case), there appears an average over the binding energy of the dimer at *infinite* separation $B_D(\infty)$,[†] i.e., the two ions are at very large separation. This condition is independent of the average ion–ion separation in the real solution. We now write the binding energy of the dimer at infinite separation in the form

$$B_D(\infty) = B_A + B_B \qquad (3.211)$$

Hence the solvation Gibbs energy may be rewritten as

$$\begin{aligned}
\Delta G_D^* &= -kT\ln\langle\exp[-\beta B_D(\infty)]\rangle_* \\
&= -kT\ln\langle\exp(-\beta B_A)\exp(-\beta B_B)\rangle_* \\
&= -kT\ln\langle\exp(-\beta B_A)\rangle_* - kT\ln\langle\exp(-\beta B_B)\rangle_* \\
&= \Delta G_A^* + \Delta G_B^* \qquad (3.212)
\end{aligned}$$

In the third equality on the rhs of equation (3.212) we exploit the fact that we have an average of a product of two functions, each of which depends on the binding energy of a single ion. These two functions, viewed as a random function defined on the configurational space of all the molecules of the system (excluding the solvaton), are independent because of their large separation. Hence the average of the product of the two functions may be factorized into a product of two averages. Finally, we identify in relation (3.212) the two solvation Gibbs energies pertaining to the two ions A and B. Thus we have here a rigorous basis for splitting ΔG_D^* into single ion contributions.

It should be stressed that the split in relation (3.212) is valid for *any* solute concentration (note the symbol $\langle\ \ \rangle_*$ defined in Section 3.12). The reason is that the factorization in relation (3.212) follows from the infinite separation of the two groups A and B of the solvaton, and *not* from the average ion–ion separation in the system. Such an argument clearly cannot be extracted from thermodynamics.

From relation (3.212) it also follows that a similar single-ion-contribution expression may be written for any other thermodynamic quantity of solvation. The argument is obviously valid also for the more general case of a multi-ionic solute.

Although we have shown that ΔG_D^* can be written rigorously as a sum of two ionic contributions, we cannot offer any means of computing each

[†] In Section 3.12 we derived an expression for ΔA_{AB}^* but the same expression holds for ΔG_{AB}^*.

of the single ion solvation Gibbs energies. This difficulty is of the same nature encountered in the conventional thermodynamical treatment. However, the structure of the statistical mechanical expressions for ΔG_A^* and ΔG_B^* might provide a better guide of how to choose a judicious approximation for calculating ΔG_A^* and ΔG_B^* from experimental results. In the course of the development of the conventional single ion solvation thermodynamics, several approximate procedures have been suggested. These have been reviewed recently.[59] We shall now show that the statistical mechanical expression for the solvation Gibbs energy does favor one of these approximations, the one suggested originally by Grunwald.[60,61]

Let us consider a bulky ion of the type depicted in Figure 3.1. We assume that the electric charge is concentrated at the center of the ion. An example of such an ion might be tetraphenylammonium ion. The binding energy of this ion can be expressed in the form

$$B_A = \sum_{i=1}^{N_w} U(A, i) = \sum_{i=1}^{N_w} \sum_{k=1}^{4} U(k, i) + \sum_{i=1}^{N_w} U(N^+, i) \qquad (3.213)$$

In this relation, the binding energy of the solute A to the solvent (here, for simplicity, we assume that the solvent consists of N_w water molecules, but all the subsequent arguments are valid for any mixture of

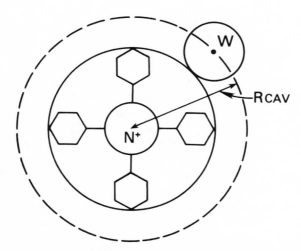

Figure 3.1. Schematic description of the excluded volume produced by a tetraalkylammonium ion in water. Obviously, in reality neither of the molecules is a sphere.

solvents and any concentration of the solute) is written as a sum of the binding energies of all the groups contained in A. First we have the four phenyl groups, and the second term on the rhs of relation (3.213) is the binding energy of the nitrogen ion to the solvent. The exact way of splitting B_A into group contributions is of no importance; as we shall see, our conclusion will be independent of such details.

We now observe the way in which the binding energy is featured in the expression for the solvation Gibbs energy of A. For simplicity we assume that A is very dilute in the solvent w. Hence

$$\Delta G_A^{*\circ} = - kT \ln \langle \exp(- \beta B_A) \rangle_0$$
$$= - kT \ln \langle \exp(- \beta B_{ph}) \exp(- \beta B_{N^+}) \rangle_0 \qquad (3.214)$$

where B_{ph} and B_{N^+} are the binding energies that correspond to the two terms on the rhs of relation (3.213), i.e., the phenyl groups and the nitrogen ion, respectively.

The average denoted by $\langle \ \ \rangle_0$ is over *all* possible configurations of the solvent molecules. This includes configurations for which some solvent molecules penetrate into the space occupied by the solute. However, for each configuration for which at least one solvent molecule penetrates the region occupied by the phenyl groups, B_{ph} becomes infinitely large, so $\exp[- \beta B_{ph}]$ becomes practically zero. In Figure 3.1 we indicate this region by the dashed line, which marks what may be called the excluded region for the solvent.

Hence the average quantity in relation (3.214) receives nonzero contributions only from those configurations for which no solvent molecules penetrate this excluded volume. This means that interaction between the charge at the central atom and any solvent molecule will be realized only at a distance of at least R_{cav} (Figure 3.1). Clearly, the bulkier the groups surrounding the charged atom (say, phenyl, alkyl phenyl, or poly-substituted phenyl groups), the larger the distance that a solvent molecule can "sense" the central charge. Thus for sufficiently large groups the factor $\exp(- \beta B_{N^+})$ in relation (3.214) will have no effect on the value of the average. From the mathematical point of view we can discard this factor without affecting the value of the integral. (Of course, from the physical, or rather the chemical, point of view the central atom holds the four phenyl groups together. However, in the theoretical expression we may fix the location of the central atom as well as the phenyl groups and need not worry about holding the groups together by chemical bonds.) Expression (3.214) may therefore be written as

$$\Delta G_A^{*\circ} \approx - kT \ln \langle \exp(- \beta B_{ph}) \rangle_0 \qquad (3.215)$$

where it is understood that the four phenyl groups in relation (3.215) hold exactly the same configuration as in relation (3.214). The approximation in relation (3.215) is better, the larger the bulky groups in the ion A.

Equation (3.215), although a good approximation for $\Delta G_A^{*\circ}$, is not practical. We do not have a "molecule" of the type depicted in Figure 3.1, but lacking its central atom. However, using a similar argument employed above, we can, instead of eliminating the factor $\exp(-\beta B_{N^+})$, replace it by a factor of the form $\exp(-\beta B_c)$, where physically we replace the charged nitrogen by an uncharged carbon atom. This replacement should have almost no effect on the average in relation (3.214); therefore we write

$$\Delta G_A^{*\circ} \approx -kT \ln\langle \exp(-\beta B_{ph}) \exp(-\beta B_c) \rangle_0$$

$$= -kT \ln\langle \exp(-\beta B_{\bar{A}}) \rangle_0 = \Delta G_{\bar{A}}^{*} \qquad (3.216)$$

where $B_{\bar{A}}$ is the binding energy of a tetraphenylmethane molecule ($\bar{A}$), a *real* molecule, for which $\Delta G_{\bar{A}}^{*}$ may be measured. Thus we arrive at the approximate relation

$$\Delta G_A^{*\circ} \approx \Delta G_{\bar{A}}^{*\circ} \qquad (3.217)$$

which, in principle, can be improved indefinitely by taking larger and larger groups instead of the phenyl radicals.

The anion B may be treated similarly. We find a neutral molecule $\bar{B}$, which has almost the same structure as B but lacks the negative charge. An example could be tetraphenylmethane for tetraphenylborate ion. Therefore using the same type of argument as above, we can approximate the solvation Gibbs energy of the salt AB by

$$\Delta G_{AB}^{*\circ} = \Delta G_A^{*\circ} + \Delta G_B^{*\circ} \approx \Delta G_{\bar{A}}^{*\circ} + \Delta G_{\bar{B}}^{*\circ} \qquad (3.218)$$

where, on the rhs, we have two independently measurable quantities, each of which represents an approximation to the solvation Gibbs energy of the anion and cation, respectively.

Finally we note that in his original work Born[62] proposed a very simple model to estimate the solvation Gibbs energy of ions. In this model a conducting sphere is charged in a vacuum and in the liquid. Although it was not stated explicitly in his article, Born does implicitly assume that the spheres are devoid of translational degrees of freedom. Therefore his computed quantities are, in principle, approximate values for ΔG_i^{*} and not for the conventional standard Gibbs energy of solution.

3.14. CALCULATION OF SOLVATION GIBBS ENERGIES FROM STANDARD GIBBS ENERGIES AND ENTHALPIES OF FORMATION OF IONIC SOLUTIONS

We consider as a specific example the case of KCl. The standard Gibbs energies and enthalpies of formation of a hypothetical 1 molal ideal solution are given as 298.15 K.[45] These are as follows (g, l, and s stand for gas, liquid, and solid respectively and NBS notation[45] is employed):

(1) $\Delta_f G^\circ(\text{KCl}, 1\text{ m}) = \mu^l(\text{KCl}, 1\text{ m}) - \mu^s(\text{K}) - \frac{1}{2}\mu^g(\text{Cl}_2, 1\text{ atm})$

(2) $\Delta_f G^\circ(\text{K}, 1\text{ atm}) = \mu^g(\text{K}, 1\text{ atm}) - \mu^s(\text{K})$

(3) $\Delta_f G^\circ(\text{Cl}, 1\text{ atm}) = \mu^g(\text{Cl}, 1\text{ atm}) - \frac{1}{2}\mu^g(\text{Cl}_2, 1\text{ atm})$

(4) $\Delta_f G^\circ(\text{K}^+, 1\text{ atm}) = \mu^g(\text{K}^+, 1\text{ atm}) - \mu^s(\text{K})$

(5) $\Delta_f G^\circ(\text{Cl}^-, 1\text{ atm}) = \mu^g(\text{Cl}^-, 1\text{ atm}) - \frac{1}{2}\mu^g(\text{Cl}_2, 1\text{ atm})$ (3.219)

A combination of $(1) - (4) - (5)$ could have yielded the standard Gibbs energy of solution. However, (4) and (5) are not available for the Gibbs energies. Instead, we take the following combinations:

$$(1) - (2) - (3) = \mu^l(\text{KCl}, 1\text{ m}) - \mu^g(\text{K}, 1\text{ atm}) - \mu^g(\text{Cl}, 1\text{ atm})$$

$$= \mu^l(\text{KCl}, 1\text{ m}) - \mu^g(\text{K}^+, 1\text{ atm}) - \mu^g(\text{Cl}^-, 1\text{ atm})$$

$$+ [\mu^g(\text{K}^+, 1\text{ atm}) + \mu^g(\text{Cl}^-, 1\text{ atm}) - \mu^g(\text{K}, 1\text{ atm})$$

$$- \mu^g(\text{Cl}, 1\text{ atm})]$$

$$= W(\text{K}^+, \text{Cl}^- \mid 0) + RT \ln \rho_{\text{K}^+}^l \rho_{\text{Cl}^-}^l \Lambda_{\text{K}^+}^3 \Lambda_{\text{Cl}^-}^3$$

$$- RT \ln \rho_{\text{K}^+}^g \rho_{\text{Cl}^-}^g \Lambda_{\text{K}^+}^3 \Lambda_{\text{Cl}^-}^3 + \Delta E(\text{ionization})$$ (3.220)

where $W(\text{K}^+, \text{Cl}^- \mid 0)$ is the solvation Gibbs energy of the ions in a pure solvent and $\Delta E(\text{ionization})$ is the Gibbs energy change for the transfer of an electron from K to Cl to form the ions K^+ and Cl^-. Since this process is carried out at the same standard states for all the species, it involves only the energy of the transfer. This may be obtained from the standard enthalpy of formation at $T = 0$ K. Hence we write

$$\Delta E(\text{ionization}) = \mu^g(\text{K}^+, 1\text{ atm}) + \mu^g(\text{Cl}^-, 1\text{ atm})$$

$$- \mu^g(\text{K}, 1\text{ atm}) - \mu^g(\text{Cl}, 1\text{ atm})$$

$$= \Delta_f H_0^\circ(\text{K}^+) + \Delta_f H_0^\circ(\text{Cl}^-) - \Delta_f H_0^\circ(\text{K}) - \Delta_f H_0^\circ(\text{Cl})$$
    (3.221)

The required quantity is thus obtained from relations (3.220) and (3.221) in the form

$$W(K^+, Cl^- \mid 0) = \Delta G^*_{solv}$$
$$= \Delta_f G^\circ(KCl, 1\ m) - \Delta_f G^\circ(K, 1\ atm)$$
$$- \Delta_f G^\circ(Cl, 1\ atm) - \Delta E(\text{ionization})$$
$$+ RT \ln \left(\frac{\rho^g_{K^+} \rho^g_{Cl^-}}{\rho^l_{K^+} \rho^l_{Cl^-}} \right) \tag{3.222}$$

where $\Delta E(\text{ionization})$ is taken from equation (3.221) and the concentrations of the ions in their standard states are given by

$$\rho^g_{K^+} = \rho^g_{Cl^-} = P/RT = 4.087 \times 10^{-2}\ \text{mol dm}^{-3} \tag{3.223}$$

and

$$\rho^l_{K^+} = \rho^l_{Cl^-} = m d^{25}_w = 0.99705\ \text{mol dm}^{-3} \tag{3.224}$$

d^{25}_w being the density of water at 25 °C. Thus for KCl we have (at 298.15 K)

$$\Delta E(\text{ionization}) = 70.19\ \text{kJ mol}^{-1} \tag{3.225}$$

$$\Delta_f G^\circ(KCl, 1\ m) - \Delta_f G^\circ(K, 1\ atm) - \Delta_f G^\circ(Cl, 1\ atm)$$
$$= -580.76\ \text{kJ mol}^{-1} \tag{3.226}$$

and

$$RT \ln \left(\frac{\rho^g_{K^+} \rho^g_{Cl^-}}{\rho^l_{K^+} \rho^l_{Cl^-}} \right) = -15.837\ \text{kJ mol} \tag{3.227}$$

Hence

$$\Delta G^{*\circ}_{KCl} = W(K^+, Cl^- \mid 0) = -580.76 - 70.19 - 15.837$$
$$= -666.8\ \text{kJ mol}^{-1} \tag{3.228}$$

Similarly, the enthalpy of solvation in pure solvent may be obtained from

$$\Delta H^{*\circ}_{KCl} = \Delta_f H^\circ(KCl, 1\ m) - \Delta_f H^\circ(K^+, 1\ atm)$$
$$- \Delta_f H^\circ(Cl^-, 1\ atm) - 2RT^2 \alpha_p + 2RT \tag{3.229}$$

where α_p is the thermal expansibility of pure water.
    For KCl we have

$$\Delta H^{*\circ}_{KCl} = -419.53 - 514.26 + 233.13 + 2 \times 2.289$$
$$= -696.08\ \text{kJ mol}^{-1} \tag{3.230}$$

From $\Delta G^{*\circ}_{KCl}$ and $\Delta H^{*\circ}_{KCl}$ we may compute the corresponding entropy of solvation in the form

$$\Delta S^{*\circ}_{KCl} = (\Delta H^{*\circ}_{KCl} - \Delta G^{*\circ}_{KCl})/T \qquad (3.231)$$

which for KCl is given by

$$\Delta S^{*\circ}_{KCl} = 98.21 \text{ J mol}^{-1} \text{K}^{-1} \qquad (3.232)$$

3.15. CONVERSION OF CONVENTIONAL STANDARD GIBBS ENERGIES AND MEAN IONIC ACTIVITIES OF 1:1 ELECTROLYTES INTO SOLVATION QUANTITIES

The expression for the CP of the solute D in a solvent l at any concentration is given in Section 3.12. If c is any concentration unit (such as molality or mole fraction), then the general conversion relation into number densities assumes the form (see also Section 3.20)

$$\rho_i = T^c_i c_i \qquad (3.233)$$

where T^c_i is the conversion factor from c_i into ρ_i for the solute i.

The CP of D in the liquid is written in two forms:

$$\mu^l_D = W(A, B \mid *) + kT \ln T^c_A T^c_B c_A c_B \Lambda^3_A \Lambda^3_B$$
$$= (\mu^{\circ c}_D + kT \ln c_A c_B + 2kT \ln \gamma^c_{AB}) \qquad (3.234)$$

and in the limit of dilute solutions

$$\mu^l_D = W(A, B \mid 0) + kT \ln T^{c,0}_A T^{c,0}_B c_A c_B \Lambda^3_A \Lambda^3_B$$
$$= [\mu^{\circ c}_D + kT \ln c_A c_B] \qquad (3.235)$$

where the expression in square brackets is the conventional thermodynamic form, and

$$T^{c,0}_i = \lim_{c_i \to 0} T^c_i \qquad (3.236)$$

The standard CP in the c scale is identified from equation (3.235) as

$$\mu^{\circ c}_D = W(A, B \mid 0) + kT \ln T^{c,0}_A T^{c,0}_B \Lambda^3_A \Lambda^3_B$$
$$= \mu^{\circ \rho}_D + kT \ln(T^{c,0}_A T^{c,0}_B) \qquad (3.237)$$

where

$$\mu_D^{\circ\rho} = W(A, B \,|\, 0) + kT \ln \Lambda_A^3 \Lambda_B^3 \tag{3.238}$$

The corresponding mean activity coefficient in the c scale is [from equations (3.234), (3.235), and (2.59)]

$$2kT \ln \gamma_{AB}^c = W(A, B \,|\, *) - W(A, B \,|\, 0) + kT \ln[(T_A^c/T_A^{c,0})(T_B^c/T_B^{c,0})]$$
$$= 2kT \ln \gamma_{AB}^\rho + kT \ln[T_A^c/T_A^{c,0})(T_B^c/T_B^{c,0})] \tag{3.239}$$

The data on the mean activity coefficient γ_{AB}^c and the conversion factors T_A^c and $T_A^{c,0}$ can be employed to obtain γ_{AB}^ρ and hence relation (2.59), which is

$$W(A, B \,|\, *) - W(A, B \,|\, 0) = 2kT \ln \rho_{AB}^\rho \tag{3.240}$$

3.16. "SOLVATION" IN ADSORPTION PROCESSES

We examine here the concepts of solvation and liberation in a very simple, and exactly solvable, model. This is the Langmuir model for adsorption. It is assumed that there are M equivalent and independent sites, and each site may accommodate at most one adsorbed molecule. For simplicity we also assume that the adsorbed molecules are structureless, and the only energy taken into account is the interaction energy U_0 between the molecule and the site.

For a system of N molecules adsorbed on N sites at temperature T, the partition function is

$$Q(N, M, T) = \frac{M!}{(M - N)! \, N!} \exp(-\beta N U_0) \tag{3.241}$$

and the CP of the adsorbed molecules is

$$\mu^{ad} = -kT \ln[Q(N + 1, M, T)/Q(N, M, T)]$$
$$= U_0 + kT \ln \left(\frac{\theta}{1 - \theta} \right) \tag{3.242}$$

where $\theta = N/M$ is the fraction of occupied sites.

The PCP is defined as the change in the Helmholtz energy of the system for the process of placing one additional molecule at a *fixed* site, i.e.,

$$\mu^{*\text{ad}} = -kT\ln[Q(N+1, M, T, \text{fixed site})/Q(N, M, T)]$$

$$= -kT\ln\left\{\frac{[(M-1)!/(M-1-N)!\ N!]\exp[-\beta(N+1)U_0]}{[M!/(M-N)!\ N!]\exp(-\beta N U_0)}\right\}$$

$$= U_0 - kT\ln(1-\theta) \tag{3.243}$$

Thus the CP in relation (3.242) can be expressed in the form

$$\mu^{\text{ad}} = \mu^{*\text{ad}} + kT\ln\theta \tag{3.244}$$

We note that here the momentum partition function does not appear, since the particles do not possess translation degrees of freedom. However, the second term on the rhs of equation (3.244) has the significance of a liberation Helmholtz energy.[†] It contains $\theta = N/M$, where M arises from the accessibility of all the M sites to the released particle [the analogue of V in equation (3.11)], and N arises, as in equation (3.11), from the assimilation of the released particle in all the N members of its species.

We now write the CP of the same molecules in an ideal-gas phase:

$$\mu^{ig} = kT\ln\rho\Lambda^3 \tag{3.245}$$

The "solvation" Helmholtz energy, defined here for the process of transferring a molecule from a *fixed point* in the gas phase to a *fixed site* on the adsorption surface, is

$$\Delta A^* = \mu^{*\text{ad}} = U_0 - kT\ln(1-\theta) \tag{3.246}$$

It consists of the interaction between the molecule and the site U_0 and a second term which is basically an entropy term, arising from the fact that placing a particle at a fixed position always lowers the entropy of the system, since there are less configurations available for the N molecules on $M-1$ sites. The entropy and energy of solvation are obtained simply from equation (3.246):

$$\Delta S^* = k\ln(1-\theta) \tag{3.247}$$

and

$$\Delta E^* = U_0 \tag{3.248}$$

[†] This is one reason why we have preferred to call this term as well as the $kT\ln\rho\Lambda^3$ term the liberation Gibbs or Helmholtz energy rather than the translational Gibbs or Helmholtz energy.

We note also that in order to measure ΔA^*, we need to consider the equilibrium condition

$$\mu^{ig} = \mu^{ad} \tag{3.249}$$

which leads to

$$\mu^{*ad} + kT \ln \theta = kT \ln \rho \Lambda^3 \tag{3.250}$$

or

$$\Delta A^* = kT \ln(\rho \Lambda^3/\theta)_{eq} \tag{3.251}$$

This is another example where, in addition to the concentrations ρ and θ at equilibrium, we also need to know the molecular quantity Λ^3. Note that the conventional standard Helmholtz energy of adsorption would have been measured by

$$\Delta A^\circ = kT \ln(\rho/\theta)_{eq} \tag{3.252}$$

and the relation between ΔA^* and ΔA° is simply

$$\Delta A^\circ = \Delta A^* - kT \ln \Lambda^3 \tag{3.253}$$

The second term on the rhs of equation (3.253) is clearly irrelevant to the study of the "solvation" of the adsorbed particle on the site.

Another noteworthy feature is that, in this very simple model, the entropy and energy of solvation are not related to each other, in contrast to cases where a compensation phenomenon exists. Here, the adsorption process has no effect on the medium (unlike, for instance, the cases treated below). The energy of solvation is simply the binding energy of the molecule to its site (environment), while the entropy of solvation arises from the restriction on the number of configurations available to the particles. This entropy tends to zero for $\theta \to 0$, but tends to $-\infty$ when $\theta \to 1$.

We now introduce a more complicated adsorption model, where the adsorbed molecules can induce "structural changes" in the sites. We shall present here only a brief description of this model and cite some of the main results. A full treatment is quite lengthy and has been published elsewhere.[30,54]

Consider again a system of M independent, identical, and localized adsorbing sites. Each site may attain one of two states L (low) and H (high) with corresponding energy levels E_L and E_H, where $E_L < E_H$. Each site can adsorb at most one gas molecule G; the adsorption energies are U_L and U_H according to whether the site is in state L or H, respectively.

For the empty sites we have equilibrium concentrations of the L and

H sites given by

$$x_L^0 = \frac{\bar{M}_L}{M} = \frac{Q_L}{Q_L + Q_H}, \qquad x_H^0 = 1 - x_L^0 \qquad (3.254)$$

where $\bar{M}_L$ is the average number of sites in the L states and Q_L and Q_H are defined by

$$Q_L = \exp(-\beta E_L) \quad \text{and} \quad Q_H = \exp(-\beta E_H) \qquad (3.255)$$

The symbol x_L^0 signifies the mole fraction of sites in state L for the empty sites, i.e., before adsorption takes place.

As a result of the different interaction energies, an adsorbed molecule G may shift the equilibrium composition of L and H, and one can compute exactly the effect of G on the equilibrium composition of L and H (details are provided elsewhere[54]). For our purposes here we cite only the results for the "solvation" Helmholtz energy, entropy, and energy of G:

$$\Delta A_G^* = -kT\ln\langle\exp(-\beta B_G)\rangle_0 - kT\ln(1-\theta)$$
$$= -kT\ln(q_L x_L^0 + q_H x_H^0) - kT\ln(1-\theta) \qquad (3.256)$$

$$\Delta S_G^* = k\ln(1-\theta) + k\ln\langle\exp(-\beta B_G)\rangle_0 + \frac{1}{T}(\langle E+B_G\rangle_G - \langle E\rangle_0) \quad (3.257)$$

$$\Delta E_G^* = \langle B_G\rangle_G + (\langle E\rangle_G - \langle E\rangle_0) \qquad (3.258)$$

where we define

$$q_L = \exp(-\beta U_L) \quad \text{and} \quad q_H = \exp(-\beta U_H) \qquad (3.259)$$

and the average $\langle\ \rangle_G$ is a conditional average, i.e., using the probabilities of finding a site in state L or state H, given that a molecule G is adsorbed on the site.

There are several points to be noted in the results (3.256) and (3.258). In the first place, the average appearing in relation (3.256) pertains to the distribution x_L^0 and x_H^0, i.e., the distribution of sites in states L and H *before* the introduction of G. The structure of the first term on the rhs of relation (3.256) is the same as the more general term given in Section 3.8.

On the other hand, both the entropy and energy of solvation on the site have a typical term, similar to the "structural changes in the solvent" that we have demonstrated in Section 3.8. Here, in this relatively simple model, the quantity $\langle E\rangle_G - \langle E\rangle_0$ expresses the change in the average energy of the site caused by the adsorption process. Thus besides $\langle B_G\rangle_G$, which is the conditional average binding energy of G to the site, we have a second contribution to ΔE_G^* which arises from the effect of G on the equilibrium composition of L and H. A similar contribution is found in ΔS_G^*. We note that these two contributions cancel each other exactly when we form the combination $\Delta E_G^* - T\Delta S_G^*$. This is why ΔA_G^* does not contain a conditional average, i.e., ΔA_G^* is determined only by the composition of the empty adsorbing sites. We see that in this model the values of ΔS_G^* and ΔE_G^* make a common contribution due to the effect of the adsorption process on the equilibrium distribution of the sites. This is in contrast to the simple adsorption model where ΔS_G^* and ΔE_G^* are completely independent of each other.

3.17. SOLVATION IN A LATTICE MODEL FOR AQUEOUS SOLUTIONS

In this section we present another, exactly solvable model where solvation phenomena might be studied. This model is somewhat more complicated than the adsorption model treated in Section 3.16. In fact, this model and a few of its variants have been studied by several authors in connection with the anomalous properties of water and aqueous soutions. Here, we give a brief description of the model and some results. A more detailed treatment has been presented elsewhere.[4]

Consider a system of N_w water molecules and N_s solute molecules, where N_L of the water molecules are presumed to form a lattice with a well-defined structure. (In some specific applications the lattice structure is identified as either the ice lattice or the clathrate lattice. Here the specific structure of the lattice will be of no interest to us.) The lattice contains empty spaces, which we shall refer to as "holes." We assume that there is only one kind of such holes and that their number is proportional to the number of lattice molcules. Thus there are altogether $N_0 N_L$ holes formed by N_L lattice molecules. The remaining $N_H = N_w - N_L$ water molecules, as well as the N_s solute molecules, are presumed to occupy these holes. We also require that each hole may accommodate at most one interstitial molecule (water or solute) and that no interaction exists between interstitial molecules in different holes.

The total energy of the system consists of $N_L E_L$ for the lattice energy, $N_H E_H$ for the total interaction energy of interstitial water molecules with the holes, and $N_s E_s$ for the total interaction energy of the solute molecules with the holes. The canonical partition function for such a system is therefore

$$Q(T, V, N_w, N_s) = \frac{(N_0 N_L)! \exp[-\beta(N_L E_L + N_H E_H + N_s E_s)]}{N_s!\, N_H!\, (N_0 N_L - N_H - N_s)!} \tag{3.260}$$

Clearly, in this particular model, fixing the volume V of the system is equivalent to fixing the number of lattice molecules N_L. The volume of the system is given by $V = N_L V_L$, where V_L is the volume per lattice molecule. The interstitial molecules do not make a direct contribution to the volume.

To allow for exchange between H and L molecules, we transform to the isothermal–isobaric partition function:

$$\varDelta(T, P, N_w, N_s) = \sum_{V=V_{min}}^{V_{max}} Q(T, V, N_w, N_s) \exp(-\beta PV)$$

$$= \sum_{N_L=N_{L,min}}^{N_w} Q(T, V, N_w, N_s) \exp(-\beta PV) \quad (3.261)$$

where we use the fact that the sum over all possible volumes is equivalent to the summation over N_L from a minimum value $N_{L,min} = (N_w + N_s)/(N_0 + 1)$ (this follows from the geometrical requirement of the model that $N_H + N_s \le N_0 N_L$) to the maximum value $N_{L,max} = N_w$.

The CP of the solute s is obtained by differentiating the Gibbs energy

$$G(T, P, N_w, N_s) = -kT \ln \varDelta(T, P, N_w, N_s) \quad (3.262)$$

i.e.,

$$\mu_s = \left(\frac{\partial G}{\partial N_s}\right)_{T,P,N_w} = E_s + kT \ln[N_s/(N_0 N_L - N_H - N_s)] \quad (3.163)$$

In equation (3.263) and the subsequent equations, N_L and N_H are the values of these variables at equilibrium.

Next, the PCP is computed. We write the partition functions Q and $\varDelta$ as before, but with the additional restriction that one s particle is at a fixed hole. This can be computed from either Q or $\varDelta$, and the result is

$$\mu_s^* = -kT \ln[Q(T, V, N_w, N_s + 1, \text{one s at a fixed hole})/Q(T, V, N_w, N_s)]$$

$$= E_s + kT \ln[N_0 N_L/(N_0 N_L - N_H - N_s)] \quad (3.264)$$

Hence from equations (3.263) and (3.264) we obtain

$$\mu_s = \mu_s^* + kT \ln \left(\frac{N_s}{N_0 N_L}\right) \quad (3.265)$$

We note that the liberation Gibbs energy, as in Section 3.16, has the assimilation term $kT \ln N_s$ and a term $kT \ln N_0 N_L$ that arises from the accessibility of all the available holes. Since in this model there are no translational degrees of freedom, no term of the kind $kT \ln \Lambda_s^3$ appears in the liberation Gibbs energy.

The solvation Gibbs energy is therefore given by

$$\Delta G_s^* = \mu_s^* = E_s + kT \ln \left(\frac{N_0 N_L}{N_0 N_L - N_H - N_s}\right) \qquad (3.266)$$

In order to compute the entropy, enthalpy, and volume of solvation, we must know how the equilibrium values of N_L and N_H [i.e., the values of N_L and N_H for which one of the terms in the sum (3.261) is maximum] change with T and P. The results are (we restrict ourselves to the case of very dilute solutions, where $N_s \ll N_L, N_H$)

$$\Delta S_s^* = -\frac{\partial \Delta G_s^*}{\partial T} = -k \ln \left[\frac{N_0 N_L}{N_0 N_L - N_H}\right] + kT \left[\frac{N_w}{N_L(N_0 N_L - N_H)}\right]\left(\frac{dN_L}{dT}\right)_{eq}$$
$$(3.267)$$

$$\Delta H_s^* = \Delta G_s^* + T\Delta S_s^* = E_s + kT^2 \left[\frac{N_w}{N_L(N_0 N_L - N_H)}\right]\left(\frac{dN_L}{dT}\right)_{eq} \qquad (3.268)$$

and

$$\Delta V_s^* = \frac{\partial \Delta G_s^*}{\partial P} = kT \left[\frac{-N_w}{N_L(N_0 N_L - N_H)}\right]\left(\frac{dN_L}{dP}\right)_{eq} \qquad (3.269)$$

where dN_L/dT and dN_L/dP are the derivatives of the average (or equilibrium) value of N_L with respect to T and P.

We have noted previously in connection with the model treated in Section 3.16 and in the more general case discussed in Section 3.8 that the solvation Gibbs energy ΔG_s^* is determined by the equilibrium values of N_L and N_H. No contribution due to structural changes in the "solvent" appears in equation (3.266). On the other hand, in ΔS_s^*, ΔH_s^*, and ΔV_s^* the terms containing the derivatives dN_L/dT and dN_L/dP may be reinterpreted as contributions due to structural changes in the solvent. A more detailed study of this model is described in Chapters 6 and 7 of Ben-Naim.[4]

3.18. SOLVATION AND HYDROPHOBIC INTERACTIONS

In this section we show that the so-called hydrophobic interaction (HI),[30] or its more general term "solvophobic interaction,"[63] is expressible in terms of differences in solvation Gibbs energies. In Section 1.2 we introduced the concept of the solvaton. The solvaton is a particular s molecule which we have chosen to place at a fixed position and to study its solvation properties. This new concept was introduced merely to distinguish between the particular s molecule in which we are interested and all other s molecules in the system. Of course, such a distinction cannot be made in practice — we cannot tag a specific molecule. However, theoretically, we *can* do it. We can always write the partition function of a system having one solvaton at some fixed position. From the point of view of the system (excluding the solvaton), this partition function is equivalent to a partition function of system subjected to an "external" field of force produced by the solvaton at a fixed position.

In most of the examples treated in this book the solvaton consists of a single entity. The only exception is the case of ionic solutes, where each solvaton consists of a pair of ions placed at fixed positions and at infinite separation from each other.

We now generalize our definition of the solvaton to include pairs, triplets, and so on, of specific molecules to which we shall refer as solvatons. As for single solvatons, these are characterized mainly by their interaction, or rather by their binding energy, with the rest of the system. However, when we have more than one solvaton, we shall require in addition that the solvatons do not interact with each other.

Thus each solvaton of species s interacts with the rest of the system in exactly the same way as an s molecule, except that it does not interact with other solvatons. Again, we stress that the solvatons cannot be physically identified. It is merely a theoretical convenience of notation when dealing with such systems.

Consider now a system containing N_A molecules of species A and N_B molecules of species B, and any other number of molecules of other species, at a given temperature T and pressure P.

We now pick up two solvatons, one A and one B molecule and define the Gibbs energy change for the process of bringing these two molecules from fixed positions at infinite separation to fixed positions at some close distance R, where R is of the order of magnitude of the mean molecular diameters of A and B, say $R \approx (\sigma_A + \sigma_B)/2$. For simplicity, we assume that A and B are spherical molecules, so that the Gibbs energy of this process is a function of R only (and not of the orientation of the two molecules).

For classical systems it is easily shown[30] that the Gibbs energy change may be written as

$$\Delta G_{AB}(R) = U_{AB}(R) + \delta G_{AB}(R) \tag{3.270}$$

where $U_{AB}(R)$ is the direct pair potential at distance R and $\delta G_{AB}(R)$ is the indirect part of the Gibbs energy change. Clearly, if we are interested in comparing the same process in different solvents, we should focus our attention on $\delta G_{AB}(R)$. The pair potential $U_{AB}(R)$ depends on the properties of the *pair* of molecules A and B and is irrelevant to the study of the solvent properties. The last argument was the main reason for referring to $\delta G_{AB}(R)$, in water, as the HI between A and B.[64] This function has also the convenient property of being independent of the direct pair potential $U_{AB}(R)$. This property has led to a useful approximate measure of the strength of the HI.[69] We note, however, that if the system contains more than one A and one B, then $\delta G_{AB}(R)$ does depend on the function $U_{AB}(R)$, but it does not depend on the pair potential between the specific two molecules that we have considered above. It is here perhaps that the introduction of the concept of two solvatons is useful. Once we declare a specific pair of molecules A and B to be solvatons, then by definition the direct interaction between the two solvatons does not exist; hence, the total Gibbs energy $\Delta G_{AB}(R)$ and the indirect part of the Gibbs energy change $\delta G_{AB}(R)$ become identical for that pair of solvatons.

It has been shown that $\delta G_{AB}(R)$ may be related to the difference in the solvation Gibbs energy of the pair of solvatons at two states,[64] at a distance R and at infinite distance, namely

$$\delta G_{AB}(R) = -kT \ln \left\{ \frac{\langle \exp[-\beta B_{AB}(R)] \rangle_*}{\langle \exp[-\beta B_{AB}(\infty)] \rangle_*} \right\}$$

$$= \Delta G_{AB}^*(R) - \Delta G_{AB}^*(\infty) \tag{3.271}$$

where $\Delta G_{AB}^*(R)$ and $\Delta G_{AB}^*(\infty)$ are the solvation Gibbs energies for the pair of solvatons at the separation R and ∞, respectively. We note also that the average $\langle \ \rangle_*$ is over all configurations of all the particles of the system, including any As and Bs but excluding the solvatons A and B.

Equation (3.271) may be simplified further by noting that the binding energy of A and B at infinite separation may be viewed as the sum of the binding energies of A and B separately. Hence we may write

$$-kT \ln \langle \exp[-\beta B_{AB}(\infty)] \rangle_*$$

$$= -kT \ln \langle \exp(-\beta B_A) \rangle_* - kT \ln \langle \exp(-\beta B_B) \rangle_*$$

$$= \Delta G_A^* + \Delta G_B^* \tag{3.272}$$

Therefore

$$\delta G_{AB}(R) = \Delta G_{AB}^*(R) - \Delta G_A^* - \Delta G_B^* \tag{3.273}$$

In equation (3.273) ΔG_A^* and ΔG_B^* are measurable quantities, while $\Delta G_{AB}^*(R)$ is not measurable. There exists no real "molecule" of the type

"A and B at distance R." However, the statistical mechanical form of $\delta G_{AB}(R)$ in equation (3.271) has suggested an approximate version of relation (3.273) in which δG_{AB} is expressible in terms of measurable quantities.[30] This aspect will not be dealt with in this section.

The quantity $\delta G_{AB}(R)$ is also related to the solvaton–solvaton distance distribution. The pair correlation function $g_{AB}(R)$ is related to the Gibbs energy change $\Delta G_{AB}(R)$ by[30]

$$
\begin{aligned}
g_{AB}(R) &= \exp[-\beta \Delta G_{AB}(R)] \\
&= \exp[-\beta U_{AB}(R)]\exp[-\beta \delta G_{AB}(R)]
\end{aligned}
\tag{3.274}
$$

Thus for any *real* pair of A and B, the pair correlation function depends on both the direct and indirect parts of $\Delta G_{AB}(R)$. However, if we focus on a pair of solvatons, then $U_{AB} = 0$ and, therefore, their pair correlation function reduces to

$$
g_{AB}(R) = \exp[-\beta \delta G_{AB}(R)]
\tag{3.275}
$$

This function is sometimes denoted by $y_{AB}(R)$.

A particular case is a pair of hard spheres (HSs) of, say, diameter σ_A in a solvent consisting of HSs with diameter σ_B. In this case the solvaton–solvaton pair correlation function is identical to the cavity–cavity pair correlation function. A pair of HS solvatons at distance R is equivalent to a pair of cavities of radius $(\sigma_A + \sigma_B)/2$ at distance R. However, for reasons given in Sections 2.2 and 3.4, the concept of cavity, in its geometric sense, is valid only in special cases of HS systems. If we have a mixture of HSs or hard particles of different shapes, then the equivalency between a HS and a cavity does not exist. Furthermore, in the case of real molecules (not hard particles), the binding energy includes both repulsive and attractive parts. For such a case the reference to $\exp[-\beta \delta G_{AB}(R)]$ as a cavity–cavity distribution is totally unjustified. In an earlier publication,[64] we used the term "source of field of force" to stress the fact that the binding energy of, say, a molecule A causes more than a cavity in the solvent. For this reason we prefer to refer to $g_{AB}(R)$ in equation (3.275) as the solvaton–solvaton pair correlation function.

Another aspect of the HI, which is intimately related with the solvation thermodynamic quantities, has been referred to recently as intramolecular HI.[65] We shall describe only briefly the ideas underlying this relationship.

Consider a "carrier" molecule, which could be a benzene, naphthalene, or even a protein molecule. Suppose that this molecule has two isomeric forms, as indicated in Figure 3.2. In one, the carrier C contains

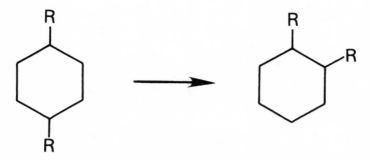

Figure 3.2. 1,4-dialkylbenzene is transformed into 1,2-dialkylbenzene. The indirect part of the Gibbs energy change for this reaction is $\delta G(4 \to 2)$.

two alkyl groups at a relatively large distance from each other, and in the other, the two alkyl groups are close to each other.

Again, as in equation (3.270), we may write the Gibbs energy change of transferring an alkyl group, say from position 4 to position 2 in a dialkylbenzene molecule, as

$$\Delta G(4 \to 2) = \Delta U(4 \to 2) + \delta G(4 \to 2) \tag{3.276}$$

where the two terms on the rhs have significance similar to that in equation (3.270). For the indirect part of the Gibbs energy change, to which we have referred as an intramolecular HI (assuming that water is the solvent), we have the following equality:[30,65]

$$\delta G(4 \to 2) = \Delta G_{1,2}^* - \Delta G_{1,4}^* \tag{3.277}$$

where $\Delta G_{1,2}^*$ and $\Delta G_{1,4}^*$ are the solvation Gibbs energies of the two isomers, which in this particular example is the 1,2-dialkylbenzene and the 1,4-dialkylbenzene. Relation (3.277) is another useful connection between a quantity conveying information on HI and solvation Gibbs energies. Clearly, by taking derivatives with respect to temperature and pressure, we can obtain the analogues of equation (3.273) or (3.277) for any other thermodynamic quantities, such as the entropy, enthalpy, volume, and so on, of the HI.

Similarly, if we look at the process of cyclization, say of n-hexane into cyclohexane, we may write the approximate relation for the indirect part of the Gibbs energy change as

$$\delta G(\text{cyclization}) \approx \Delta G_{\text{cyclohexane}}^* - \Delta G_{n\text{-hexane}}^* \tag{3.278}$$

Again, here we have a useful relation between a measure of intramolecular hydrophobic interaction and a difference between two solvation Gibbs energies.

3.19. DISTRIBUTION OF HYDROPHOBIC REGIONS ON PROTEINS

This section is an extension of both Sections 3.11 and 3.18. It extends the treatment of Section 3.11 from the problem of preferential solvation of simple solvatons to the case of complex molecules, such as a protein. The essential new feature of the protein versus the simple solvaton is that we can ask more detailed questions regarding the preferential solvation at a specific *site* or *region* on the protein. It also extends the treatment of Section 3.18 to include HI between large molecules, such as protein–protein or protein–ligand. In addition, we can define the HI between a ligand and a specific site on the protein. This will lead to a classification of the various sites (or regions) according to their HI with a test molecule, or their "hydrophobicity."

Consider, for simplicity, a single solvaton C, which is a macromolecule with an inhomogeneous surface. The solvent consists of N_w water molecules w and one organic molecule A which has a hydrophobic part, e.g., n-alkanol or a fatty acid.

We now divide the surface of C into binding sites or regions, each of which may bind either a water or an A molecule. The method of dividing the surface into sites is of no importance. In this system, all the sites will be occupied by water molecules except for the possibility of one site that is occupied by the A molecule. We denote by C_i the ith site and by AC_i the protein with its ith site occupied by A (or the protein–A complex at site i). See Figure 3.3.

The probability ratio of finding the ligand A on sites C_i and C_j is given by

$$\frac{P(AC_i)}{P(AC_j)} = \frac{\int dV \int d\mathbf{X}^{N_w} \exp\{-\beta[PV + U_{N_w} + B(AC_i) + U_{AC_i}]\}}{\int dV \int d\mathbf{X}^{N_w} \exp\{-\beta[PV + U_{N_w} + B(AC_j) + U_{AC_j}]\}} \quad (3.279)$$

where U_{N_w} is the total interaction energy among the solvent molecules; $B(AC_i)$ is the binding energy of the protein–A complex, at site C_i, to the rest of the system, and U_{AC_i} is the direct interaction between A and site C_i. The integrations are over all possible configurations of the solvent molecules and over all possible volumes of the system (at given T and P).

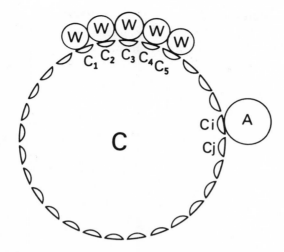

Figure 3.3. Schematic illustration of binding of a "test molecule" A to a site C_i on a macromolecule C. All other sites are presumed to be bound to water molecules.

Since U_{AC_i} depends only on the coordinates of A and C_i, the factor $\exp(-\beta U_{AC_i})$ may be factored out from the integral. Also, by dividing both the numerator and denominator by the factor

$$\int dV \int d\mathbf{X}^{N_w} \exp(-\beta PV - \beta U_{N_w}) \qquad (3.280)$$

we may rewrite expression (3.279) in the form

$$\frac{P(AC_i)}{P(AC_j)} = \frac{\exp(-\beta U_{AC_i})}{\exp(-\beta U_{AC_j})} \frac{\langle \exp[-\beta B(AC_i)]\rangle_0}{\langle \exp[-\beta B(AC_j)]\rangle_0}$$

$$= \exp[-\beta \Delta U(j \to i)] \exp[-\beta \delta G(j \to i)] \qquad (3.281)$$

where the symbol $\langle \quad \rangle_0$ denotes the average over all the volumes and all the configurations of the solvent molecules; $\Delta U(j \to i)$ is the direct interaction energy required to transfer A from C_j to C_i, and $\delta G(j \to i)$ is the indirect or the solvent-induced part of the Gibbs energy change for transferring A from C_j to C_i.[†]

[†] Note that since C is in almost pure water, all the sites are in contact with water except for the case when A occupies a given site.

The probability ratio is clearly determined by both the direct and indirect parts of the Gibbs energy change. However, if we are interested *only* in the solvent-induced contribution, we should focus on the second factor on the rhs of relation (3.281). This is equivalent to a declaration of C and A as a pair of solvatons. We rewrite the probability ratio for the case of two solvatons as

$$\frac{Y(AC_i)}{Y(AC_j)} = \exp[-\beta\delta G(j \to i)] = \exp[-\beta(\Delta G^*_{AC_i} - \Delta G^*_{AC_j})] \qquad (3.282)$$

where $\Delta G^*_{AC_i}$ and $\Delta G^*_{AC_j}$ are the solvation Gibbs energies of the complexes AC_i and AC_j, respectively.

At this point it is worth remarking on a common misconception, or rather misinterpretation, which is often found in biochemical literature. When a hydrophobic solute A (a ligand or substrate) binds on a protein (such as an enzyme), one commonly finds the argument that the binding is governed by "hydrophobic interaction." Yet, at the same time, it is explained that the HI is determined by the van der Waals forces between the solute and the site. From the above analysis, it is quite clear that, although *both* the direct (e.g., van der Waals) and indirect (i.e., HI) interaction play a role in the binding process, the two parts have different origins. The first results from the properties of the two binding molecules (the solute A and the protein), while the second results from the presence of the solvent molecules. One cannot "explain" the HI between a ligand and a site in terms of the direct van der Waals interaction between these two partners.

In order to further analyze the ratio (3.282), we assume that the binding energy of the complex AC_i may be expressed as a sum over all interactions between the groups C_i on C and all the water molecules, i.e.,

$$B(AC_i) = \sum_{k=1}^{N_w} \left[U(\mathbf{R}_A, \mathbf{X}_k) + \sum_{l \neq i} U(\mathbf{R}_l, \mathbf{X}_k) \right] \qquad (3.283)$$

Here, $U(\mathbf{R}_A, \mathbf{X}_k)$ is the interaction energy between A at $\mathbf{R}_A$ and the kth solvent molecule being at the configuration $\mathbf{X}_k$, and $U(\mathbf{R}_l, \mathbf{X}_k)$ is the interaction energy between the site C_l and the kth solvent molecule. We note that the sum over l excludes the site C_i. This follows from the assumption that C_i is "covered" by A and is thus not exposed to interaction with solvent molecules (Figure 3.3). The missing term in the sum (3.283) is exactly the binding energy of the site C_i, namely

$$B(C_i) = \sum_{k=1}^{N_w} U(\mathbf{R}_i, \mathbf{X}_k) \qquad (3.284)$$

We now rewrite equation (3.283) in the form

$$B(AC_i) = B(A) + \sum_{l \neq i} B(C_l) + B(C_i) - B(C_i)$$

$$= B(A) + \sum_{l} B(C_l) - B(C_i) \qquad (3.285)$$

and substitute in equation (3.282) to obtain

$$\frac{Y(AC_i)}{Y(AC_j)} = \frac{\left\langle \exp\left\{-\beta\left[B(A) + \sum_{l} B(C_l) - B(C_i)\right]\right\}\right\rangle_0}{\left\langle \exp\left\{-\beta\left[B(A) + \sum_{l} B(C_l) - B(C_j)\right]\right\}\right\rangle_0}$$

$$= \frac{\langle g \exp[+\beta B(C_i)]\rangle_0}{\langle g \exp[+\beta B(C_j)]\rangle_0} \qquad (3.286)$$

where the factor $g > 0$ is independent of the index i. The normalization condition

$$\sum_{i} P(AC_i) = 1 \qquad (3.287)$$

is now used to obtain the probability $P(AC_i)$ for the binding of A onto the site C_i,

$$P(AC_i) = \frac{1}{N} \exp(-\beta U_{AC_i})\langle g \exp[+\beta B(C_i)]\rangle_0 \qquad (3.288)$$

where N is a normalization constant. The conclusion from equation (3.288) is the following. When a nonpolar solute (or hydrophobic ligand) A approaches a protein, it might "land" on various sites. The direct interaction $U(AC_i)$ will favor binding to the site that exerts the larger attractive forces on A. On the other hand, the indirect (or HI) part will give preference to that site which has the weaker binding Gibbs energy to the *solvent*. The ultimate decision on which site to bind is, of course, a result of the combined contributions of the direct and indirect parts. The larger the quantity $Y(AC_i)$, the larger the solvent-induced contribution to binding on C_i. We can say that the relative hydrophobicity of the various sites is measured by the quantity $Y(AC_i)$. It is important to stress that the "hydrophobicity" of the site C_i is *not* characterized by the van der Waals forces (or any other direct forces) between A and C_i, but by the strength of the interaction of the site C_i to the *solvent* molecules.

The conclusion reached above is, of course, consistent with the experimental finding that the most hydrophobic regions on the surface of a protein are usually those that contain nonpolar groups, i.e., that have relatively weak interaction with the water molecules.

Until now we have defined the concept of the relative hydrophobicity of various sites (or regions) on a surface of a macromolecule and characterized such regions in terms of the average binding Gibbs energy of the sites toward the solvent. We now turn to the question of the measurability of the hydrophobicity of various sites. This is admittedly a difficult question and, to the best of the author's knowledge, no one has ever suggested such an experimental procedure. We outline here two possible routes along which one might answer such a question by experimental means.

One method could be a generalization of the method of measuring intramolecular HI, discussed in Section 3.18. In fact, equation (3.282) is very similar to equation (3.277). Here, we are not interested in bringing two alkyl groups to a close distance, but we wish to measure the relative hydrophobicity of the various sites C_i. Thus instead of transferring A from a site C_j to a site C_i, we consider the process of bringing the solvaton A from a fixed position at infinite separation from C to a fixed position on the site C_i. The indirect part of the Gibbs energy change is

$$\delta G(\infty \to C_i) = \Delta G_{AC_i}^* - \Delta G_C^* - \Delta G_A^* \tag{3.289}$$

where ΔG_A^* and ΔG_C^* are the solvation Gibbs energies of A and C, respectively, and in principle are measurable quantities. The more difficult quantity to measure is $\Delta G_{AC_i}^*$, i.e., the solvation Gibbs energy of the complex AC_i. (If there is only one binding site, then $\Delta G_{AC_i}^*$ could be measured, but in this case the whole problem of the relative hydrophobicity does not arise.) If we could have measured $\Delta G_{AC_i}^*$ and $\Delta G_{AC_j}^*$, then we could determine the ratio (3.282).

We suggest that an approximate way of estimating $\delta G(\infty \to i)$ might be possible by using model systems. Suppose we have the full information on the distribution of functional groups on the surface of the protein, such as alkyl, hydroxyl, or amine groups. Instead of estimating $\delta G(\infty \to i)$ for binding A to the site C_i, we may choose a model molecule that contains similar functional groups. See the schematic illustration in Figure 3.4. If A is, say, a methane molecule and the site C_i is characterized by a hydroxyl group OH, then $\delta G(\infty \to i)$ might be approximated by the process of bringing CH_4 to a hydroxyl group on, say, benzene. Next, we exploit the fact that δG does not contain the direct interaction between the two binding partners. We can therefore approximate the Gibbs energy of bringing methane to the hydroxyl group by the Gibbs energy of attaching a methyl group to the hydroxyl group to form a

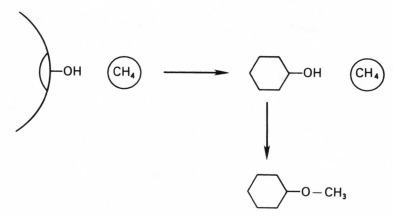

Figure 3.4. Schematic illustration of the process of attaching a methane to a hydroxyl group on a macromolecule. Instead, we can attach the same methane to a hydroxyl group on a small model compound. The indirect part of the Gibbs energy change for this process is given by the approximate relation (3.290).

methoxy group. The two approximations are

$$\delta G(\infty \rightarrow C_i) \approx \delta G(\infty \rightarrow \varphi\text{-OH}) \approx \Delta G^*_{\varphi\text{-OCH}_3} - \Delta G^*_{\varphi\text{-OH}} + \Delta G^*_{\text{CH}_4} \qquad (3.290)$$

where φ-OH and φ-OCH$_3$ stand for phenol and the methoxybenzene, respectively. All the quantities on the rhs of relation (3.290) are measurable quantities. Hence this leads to an approximate value for $\delta G(\infty \rightarrow C_i)$, which might be used to estimate the relative values of $Y(\text{AC}_i)$ through equation (3.282). We note that, although the above procedure might yield only a rough estimate for $\delta G(\infty \rightarrow C_i)$, the *relative* hydrophobicity of various sites might be predicted more accurately. The reason is that the approximations in relation (3.290) are applied more or less uniformly for all the sites, thus introducing roughly the same error in the computations of every $Y(\text{AC}_i)$.

The second experimental source that might lead to a characterization of the protein's surface in terms of the relative hydrophobicity of the various sites (or regions) is the following. Suppose the protein C is dissolved in a solution of water and a hydrophobic ligand A with density ρ_A. If the protein C contains M sites of identical binding energy toward the ligand A, then a Langmuir-type isotherm could be measured (say by some spectroscopic means):

$$\bar{N}_A = M \frac{K\rho_A}{1 + K\rho_A} \qquad (3.291)$$

where K is the binding constant and $\bar{N}_A$ the average number of ligands bound to C. From a measurement of $\bar{N}_A$ as a function of ρ_A, one can estimate both M and K. The latter is essentially a measure of the Gibbs energy of binding of A to the site.

A more complex situation arises when there are different kinds of sites. Let us assume that we have M_i sites of type C_i, which are characterized by binding constant K_i. The generalized Langmuir isotherm for this case is

$$\theta = \frac{\bar{N}_A}{M} = \sum_i x_i \frac{K_i \rho_A}{1 + K_i \rho_A} \tag{3.292}$$

where x_i is the mole fraction of sites of type C_i and θ is the total fraction of occupied sites. Here, of course, it is more difficult to estimate both x_i and K_i from a single isotherm. Some estimates of this kind for a small number of types of sites have been reported.[48] One may expect that, with accurate data on binding isotherms of the type (3.292), it will be possible to obtain a reasonable estimate of both the distribution (x_i) of the sites and their relative hydrophobicities (which are essentially contained in the quantities K_i).

3.20. SOME CONCENTRATION UNITS AND CONVERSIONS

We present here some definitions of the most common concentration units used in solution chemistry.

The mole fraction of the component i is defined by

$$x_i = n_i \bigg/ \sum_j n_j \tag{3.293}$$

where n_i is the number of molecules (or moles) of the ith species. The sum in the denominator is extended over all species present in the system. We note that in some cases there exists an ambiguity in determining the number of species (e.g., micellar soutions, dissociable solutes, and so on).

The molality of component i is usually used when i is a solute dissolved in a solvent, say w. This is defined as

$$m_i = 1000 n_i / M_w n_w \tag{3.294}$$

where n_w is the number of solvent molecules (or moles) and M_w the molecular mass or molar mass (in g mol^{-1}) of the solvent molecules.

The number density or molarity of the component i is defined by

$$\rho_i = n_i/V \tag{3.295}$$

where V is the volume of the system.

In thermodynamics, it is customary to use the following units: for m_i, moles of i per one kg of solvent; for ρ_i, moles of i in a liter (dm³) of solution.

Let d_l be the density of the solution in g cm^{-3} and d_w the corresponding density of the solvent, usually water in the solution. The following are the conversions from x_i and m_i into ρ_i.

If V is measured in cm³, then

$$d_l V = \sum_j n_j M_j \tag{3.296}$$

Hence ρ_i is given in molecules cm^{-3} (or mole cm^{-3}) by

$$\rho_i = \frac{n_i}{V} = \frac{x_i \sum_j n_j}{V} = \frac{d_l x_i}{\sum_j x_j M_j} \tag{3.297}$$

$$\rho_i = \frac{n_i}{V} = \frac{m_i M_w n_w}{1000 \, V} = \frac{d_w m_i}{1000} \tag{3.298}$$

Therefore, the conversion factors introduced in Section 3.15 are

$$T_i^x = \frac{d_l}{\sum_j x_j M_j} \tag{3.299}$$

and

$$T_i^m = \frac{d_w}{1000} \tag{3.300}$$

3.21. ESTIMATES OF THE $P\Delta V_s^*$ TERM FOR SOLVATION PROCESSES

In several places throughout the book we stated that ΔG_s^* and ΔA_s^* for the same solvation process as defined in Section 1.2 are almost identical. Likewise, the difference between ΔH_s^* and ΔE_s^* is claimed to be negligible. The exact differences between these quantities are

$$\Delta G_s^* = \Delta A_s^* + P\Delta V_s^* \tag{3.301}$$

and

$$\Delta H_s^* = \Delta E_s^* + P\Delta V_s^* \tag{3.302}$$

We now estimate the values of $P\Delta V_s^*$ for three cases.

(1) *Solvation of hexane in pure hexane* (*at 25 °C*). The molar volume of hexane is $V_m = 130.3 \text{ cm}^3 \text{ mol}^{-1}$; hence, the solvation volume of hexane in pure hexane is

$$\Delta V_s^* = V_m - RT\beta_T = 130.3 - 3.97 = 126.3 \text{ cm}^3 \text{ mol}^{-1} \tag{3.303}$$

and $P\Delta V_s^*$ at 1 atm is (1 cm^3 atm = 0.1013 J)

$$P\Delta V_s^* = 12.79 \ 10^{-3} \text{ kJ mol}^{-1} \tag{3.304}$$

The Gibbs energy of solvation of hexane in pure hexane is

$$\Delta G_s^* \approx 11.00 \text{ kJ mol}^{-1} \tag{3.305}$$

Therefore, $P\Delta V_s^*$ is quite negligible compared with ΔG_s^*.

(2) *Solvation of water in pure water* (*at 25 °C*). For pure water the molar volume is $V_m \approx 18 \text{ cm}^3 \text{ mol}^{-1}$; hence

$$\Delta V_s^* = 18 - 1.13 = 16.87 \text{ cm}^3 \text{ mol}^{-1} \tag{3.306}$$

and

$$P\Delta V_s^* = 1.71 \ 10^{-3} \text{ kJ mol}^{-1} \tag{3.307}$$

This should be compared with the solvation Gibbs energy of water in pure water given by

$$\Delta G_s^* = 26.5 \text{ kJ mol}^{-1} \tag{3.308}$$

(3) *Solvation of methane in water.* In this case we have

$$\Delta V_s^* \approx 50 \text{ cm}^3 \text{ mol}^{-1} \tag{3.309}$$

$$P\Delta V_s^* \approx 5.06 \ 10^{-3} \text{ kJ mol}^{-1} \tag{3.310}$$

and

$$\Delta G_s^* \approx 8.33 \text{ kJ mol}^{-1} \tag{3.311}$$

We note also that in Section 2.15 we used data on the solvation Gibbs energies at pressures equal to the vapor pressure of the pure hydrocarbons. However, to a good approximation the variation in pressure between 0 and 1 atm causes a negligible effect on ΔG_s^*. To see this we assume that ΔV_s^* is approximately independent of pressure; hence

$$\Delta\Delta G_s^* = \Delta G_s^*(P = 1 \text{ atm}) - \Delta G_s^*(P = 0 \text{ atm})$$

$$= \int_0^1 \Delta V_s^* dP \approx \Delta V_s^* P \tag{3.312}$$

Hence the values of $P\Delta V_s^*$, as calculated above, show that the variation of ΔG_s^* between 0 and 1 atm is quite negligible.

Finally, we note that, although ΔG_s^* and ΔA_s^* have almost the same values for the processes discussed in this book, they are *different* quantities. On the other hand, in Section 3.1 we discussed the identity of ΔG_s^* (at given T and P) and ΔA_s^* (at given T and V). These two quantities have the same value provided $\langle V \rangle$ in the T, P system is equal to V in the T, V system. Here, ΔG_s^* and ΔA_s^* refer to *different* processes of solvation and not to the same solvation process discussed in this section.

3.22. THE BEHAVIOR OF THE SOLVENT EXCESS CHEMICAL POTENTIAL IN THE LIMIT OF LOW SOLUTE CONCENTRATIONS

Let w be the solvent and s a nonionic solute. An exact result due to Kirkwood and Buff (details of the derivation of this equation are given in Chapter 4 of Ben-Naim[4]) is the following:

$$\left(\frac{\partial \mu_w}{\partial x_w}\right)_{P,T} = kT\left(\frac{1}{x_w} - \frac{\rho_s \Delta_{sw}}{1 + \rho_s x_w \Delta_{sw}}\right) \tag{3.313}$$

where x_w is the mole fraction of the solvent and Δ_{sw} is defined by

$$\Delta_{sw} = G_{ss} + G_{ww} - 2G_{sw} \tag{3.314}$$

$G_{\alpha\beta}$ are the Kirkwood–Buff integrals defined in Section 3.10.

We integrate equation (3.313) to obtain

$$\mu_w(T, P, x_s) = \mu_w^p(T, P) + kT \ln x_w + kT \int_0^{x_s} \frac{\rho_s \Delta_{sw}}{1 + \rho_s x_w \Delta_{sw}} dx_s' \tag{3.315}$$

where the constant of integration (at $x_s = 0$) is identified as the CP of pure w (note $dx_s = -dx_w$).

We now identify the excess CP of the solvent (with respect to the symmetrical ideal solutions) as

$$\mu_w^E = kT \int_0^{x_s} \frac{\rho_s \Delta_{sw}}{1 + \rho_s x_w \Delta_{sw}} \, dx_s' \tag{3.316}$$

On expanding μ_w^E to second order in the mole fraction $x_s^\dagger$ we obtain

$$\mu_w^E = \mu_w^E(x_s = 0) + x_s \frac{\partial \mu_w^E}{\partial x_s} + \tfrac{1}{2} x_s^2 \frac{\partial^2 \mu_w^E}{\partial x_s^2} + \cdots \tag{3.317}$$

The coefficients are

$$\mu_w^E(x_s = 0) = 0 \tag{3.318}$$

$$\frac{1}{kT} \frac{\partial \mu_w^E}{\partial x_s}\bigg|_{x_s=0} = \frac{\rho_s \Delta_{sw}}{1 + \rho_s x_w \Delta_{sw}}\bigg|_{x_s=0} = 0 \tag{3.319}$$

and

$$\frac{1}{kT} \frac{\partial^2 \mu_w^E}{\partial x_s^2}\bigg|_{x_s=0} = \frac{\partial}{\partial x_s} (\rho_s \Delta_{sw})|_{x_s=0} = \Delta_{sw}^0 \left(\frac{\partial \rho_s}{\partial x_s}\right)\bigg|_{x_s=0}$$

$$= \Delta_{sw}^0 \frac{\partial}{\partial x_s} (\rho_T x_s)|_{x_s=0} = \Delta_{sw}^0 \rho_T |_{x_s=0} = \Delta_{sw}^0 \rho_w^0 \tag{3.320}$$

Hence

$$\mu_w^E \approx \tfrac{1}{2} x_s^2 \frac{\partial^2 \mu_w^E}{\partial x_s^2}\bigg|_{x_s=0} = \tfrac{1}{2} x_s^2 kT \Delta_{sw}^0 \rho_w^0 \tag{3.321}$$

where the superscript zero indicates the limit of $x_s \to 0$. Thus $\mu_w^E \approx O(x_s^2)$ or $O(m_s^2)$, which justifies the considerations in Section 1.7. We note that a similar expression to (3.321) is known in the theory of regular solutions. However, in the latter it is assumed that Δ_{sw} is very small while here no such assumption has been made (for more details see Chapter 4 of Ben-Naim[4]).

---

† Note that here we are interested only in an examination of μ_w^E, not of the CP itself.

Chapter 4

On Mixing and Assimilation

4.1. INTRODUCTION

This chapter deals with mixing thermodynamics, a topic which is indirectly related to the main subject matter of this book.

Suppose we have a liquid mixture l at a given T and P and consisting of N_i molecules of species i. The total Gibbs energy of the system is

$$G^l = \sum_i N_i \mu_i$$
$$= \sum_i N_i W(i \mid l) + \sum_i N_i kT \ln \rho_i^l \Lambda_i^3 \qquad (4.1)$$

where $W(i \mid l)$ is the coupling work of i to the liquid mixture l, and for simplicity we assume that the molecules do not possess any internal degrees of freedom. The sum over i is over all the species that are present in the system.

Let μ_i^p be the chemical potential (CP) of the pure ith species at the same P and T. The total Gibbs energy of the combined pure species, before the mixing, is then

$$G^p = \sum_i N_i \mu_i^p$$
$$= \sum_i N_i W(i \mid i) + \sum_i N_i kT \ln \rho_i^i \Lambda_i^3 \qquad (4.2)$$

where $W(i \mid i)$ is the coupling work of i to a pure liquid i at the same P and T, and ρ_i^i is the density of i in this pure state. Thus the Gibbs energy change for the process of formation of the mixture from the pure liquids is given by

$$\Delta G^M = G^l - G^p = \sum_i N_i [W(i \mid l) - W(i \mid i)] + \sum_i N_i kT \ln(\rho_i^l / \rho_i^i) \qquad (4.3)$$

205

A particular case of equation (4.3) that appears in almost all text-books of thermodynamics is the case of mixing ideal gases. In this case the first sum on the rhs of equation (4.3) is zero (no interactions). Also, since the total pressure of the mixture is the same as the pressure of each pure component i before the process, we have

$$P = kT \sum \rho_i^l = kT\rho_i^i \qquad (4.4)$$

Hence

$$\rho_i^l/\rho_i^i = \rho_i^l/\sum \rho_i^l = x_i \qquad (4.5)$$

where x_i is the mole fraction of the ith species in the ideal-gas mixture. In this case equation (4.3) reduces to

$$\Delta G^M = \sum_i N_i kT \ln x_i \qquad (4.6)$$

This quantity is commonly referred to as the "free energy of mixing" (or Gibbs energy of mixing). Similarly, the quantity

$$\Delta S^M = - \sum_i N_i k \ln x_i \qquad (4.7)$$

is referred to as the "entropy of mixing." These terms are applied to quantities of the form (4.6) and (4.7) even for systems of interacting molecules. Thus in the general equation (4.3), the first term is referred to as the interaction Gibbs energy while the second term, or a modified form of it, is referred to as the mixing Gibbs energy.

Clearly, the first sum on the rhs of equation (4.3) involves the solvation Gibbs energies of the various species in the pure and mixed states, and therefore the quantity ΔG^M *is* directly relevant to the main topic of this book.

On the other hand, the so-called "Gibbs energy of mixing" terms of the form (4.6) are not directly relevant to the solvation phenomenon. However, since the interpretation of such terms does affect the interpre-tation of the remaining part in ΔG^M given by equation (4.3), we believe that the treatment of the ideal-gas mixtures is also indirectly related to the solvation phenomenon.

In Sections 4.2 to 4.8, we shall deal exclusively with ideal gases. The purpose of these sections is to examine the suitability of the terms "mixing entropy" and "mixing Gibbs energy" and the validity of the statement that "the mixing process is essentially irreversible" (see, for example, Denbigh[66]). The motivation for elaborating on this topic for *ideal-gas* mixtures arises from the realization that some faulty statements are made in many textbooks and research articles that lead to common misconceptions and often erroneous interpretations of experimental results. Of course, the more important cases are the systems of interacting particles. These systems are treated in Sections 4.9 through 4.11.

4.2. TWO SIMPLE MIXING PROCESSES

We begin by considering a classical process of mixing as described by most textbooks of thermodynamics and statistical thermodynamics. In this book we follow the presentation of Denbigh.[66] For simplicity we discuss only a two-component system (rather than a multicomponent system) and also take the simplest case as depicted in process I (Figure 4.1). In this process we have two compartments, one of which contains N_A molecules A in volume V_A, and the second N_B molecules B in volume V_B. Furthermore, for simplicity we assume that

$$N = N_A = N_B, \quad V = V_A = V_B \tag{4.8}$$

Clearly, with this specification we fulfill the conditions required by Denbigh[66]† (namely that the initial pressures P_A and P_B before the process are the same as P, the total pressure after the process).

An elementary calculation leads to the result that the Gibbs energy change in process I (Figure 4.1) is

$$\Delta G^I = N_A kT \ln x_A + N_B kT \ln x_B \tag{4.9}$$

and the corresponding entropy change is

$$\Delta S^I = - N_A k \ln x_A - N_B k \ln x_B \tag{4.10}$$

where x_A and x_B are the mole fractions of A and B, respectively. In our particular example, equations (4.9) and (4.10) reduce to

$$\Delta G^I = 2NkT \ln \tfrac{1}{2} \tag{4.11}$$

and

$$\Delta S^I = - 2Nk \ln \tfrac{1}{2} \tag{4.12}$$

The process may be carried out either adiabatically or isothermally. We

---

† The critical discussion of this particular mixing process is by no means a criticism of Denbigh's book, which I believe is an excellent book on chemical thermodynamics.

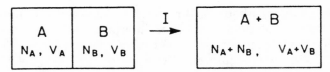

Figure 4.1. A classical process of mixing. A partition separating two compartments, each of which contains different molecules, A and B, is removed. A spontaneous mixing is observed. In this process, denoted I, we assume that the gases are ideal and that $N = N_A = N_B$ and $V = V_A = V_B$ (the process may be carried out either isothermally or adiabatically).

shall use this simplest example as representative of the general case treated by Denbigh.[66] All the conclusions of this chapter are, however, valid for more general cases also.

Having attained the results (4.9) and (4.10), Denbigh comments that these results are "in accordance with the *essential irreversibility* of *the mixing process* when it occurs at constant volume" (my italics).

Next, he proceeds to another special case, namely when the partial pressures in the mixture are all equal to the original pressure of the unmixed gases. This is shown in process II (Figure 4.2). For this process he obtains the result

$$\Delta G^{II} = 0 \qquad (4.13)$$

$$\Delta S^{II} = 0 \qquad (4.14)$$

and comments that the "changes in G and S due to the *irreversible mixing process* may be said to be *exactly cancelled* by equal and opposite effects due to the reduction in volume of the system" (my italics).

The presentation given above is typical of most textbooks on the subject, although some authors comment that the interpretation should be treated with caution.

The results of equations (4.9) and (4.10) are, of course, correct and well known. It is clear that "the essential irreversibility of the mixing process" suggests to the reader that the mixing *is* responsible for results

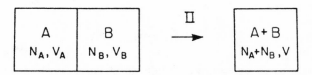

Figure 4.2. Process II. Two different gases, A and B initially in V_A and V_B, respectively, form a mixture in a final volume V. As in process I, we assume that the gases are ideal and that $N = N_A = N_B$, $V = V_A = V_B$. The process is carried out at constant temperature.

(4.9) and (4.10). Some authors refer to the "diffusion"[67] or "interdiffusion"[68] of A into B and vice versa. Furthermore, it is also clear that in process II Denbigh identified *two* contributions to ΔG^{II} (and to ΔS^{II}), namely the *mixing process* and the *compression process*, which happen to cancel each other.

Let us put the last statement more formally. Two distinct effects, namely *mixing* and *compression*, produce entropy changes ΔS^{II}_{Mix} and ΔS^{II}_{Comp}, respectively, which combine as

$$\Delta S^{II} = \Delta S^{II}_{Mix} + \Delta S^{II}_{Comp} = 0 \qquad (4.15)$$

i.e.,

$$\Delta S^{II}_{Mix} = - \Delta S^{II}_{Comp} \qquad (4.16)$$

At this point we assert that the reasoning leading to equations (4.15) and (4.16) is faulty. In process I we have only expansion, while process II may be viewed either as a combination of *two* processes, expansion and compression, or equivalently as a *nonprocess*. Furthermore, the apparent *irreversibility* of the *mixing process* I is accidental. There exist mixing processes that are reversible and *demixing* processes that are irreversible (see Section 4.4). Therefore, the characterization of the mixing processes as "essentially irreversible" is incorrect! We shall also show that terms of the form (4.10) or (4.12), which are traditionally referred to as the "mixing entropy," never arise from nor are caused by the mixing process.

Therefore, we suggest that the word "mixing," in the terms "mixing entropy" or "mixing Gibbs energy," be used only *descriptively* but never *causatively* in the sense that, although we do *observe* a mixing phenomenon, nevertheless the mixing process is never the source of what is traditionally called the "mixing Gibbs energy" and the "mixing entropy." We shall elaborate further on the meaning of the *descriptive* and *causative* parts of the process in Section 4.4.

4.3. SOME SIMPLE APPLICATIONS OF THE EXPRESSION FOR THE CHEMICAL POTENTIAL

In Section 3.1 we derived the general expression for the CP per molecule of component A, in a mixture of A and B,

$$\mu_A = W(A \mid A + B) + kT \ln \rho_A \Lambda_A^3 \qquad (4.17)$$

For simplicity, we assume that the system contains only two components A and B and that the particles are structureless. For the sake of clarity, we repeat here the interpretation of the second term on the rhs of equation (4.17), which we rewrite as

$$kT \ln \rho_A \Lambda_A^3 = kT \ln \Lambda_A^3 - kT \ln V + kT \ln N_A \qquad (4.18)$$

The first term arises from the translational momenta of the (classical) particle. The second term arises from the fact that the particle can wander in the entire volume V. The third term arises from the indistinguishability of the N_A particles. Before it was added to the system, the A particle added was *distinguishable* from the N_A particles in the system. This is so simply because this particle was *outside* our system.[†] Once introduced into the system, it loses its distinguishability. We say that this particle has been *assimilated* by the N_A A particles of the system. [As we have shown in Section 3.1, this term arises from $(N_A + 1)!/N_A!$ in the statistical mechanical expression for the CP.] We shall therefore refer to the term $kT \ln N_A$ as the assimilation Gibbs energy term.

To summarize, when we add an A particle to the system we recognize four contributions to its CP. These are the coupling work, the momentum, the volume, and the assimilation terms.

In the context of the present chapter, the momentum term will never change, so we shall focus attention on the remaining contributions to the CP. We now turn to some specific applications of equation (4.17).

(1) *A single A particle in V.* In this case, relation (4.17) reduces to

$$\mu_A = kT \ln \rho_A \Lambda_A^3 = \{\mu_A^{\circ g} + kT \ln \rho_A\} \qquad (4.19)$$

i.e., there is no coupling term and the number density is simply $\rho_A = 1/V$. In particular, we stress that the assimilation term in equation (4.19) is *zero*. In the latter equation and in subsequent equations in this section, we write in braces the same quantity but in the conventional thermodynamic notation. Here $\mu_A^{\circ g}$ is the so-called standard CP. We note also that thermodynamics alone does not provide any interpretation of the separate parts of the CP. On the other hand, statistical mechanics does indicate the origin of these terms, which in our example are the momentum and volume terms only. Obviously, it makes no sense to refer to this term as the "*mixing* Gibbs or Helmholtz energy" as is often done in the literature.[§] Clearly, a single particle wandering alone in the entire volume

---

[†] This is also true of we place A at a fixed position in the mixture. See also Section 3.1.

[§] The present author admits committing this error,[64] but other authors have done likewise.[52,53]

does not "mix" with other particles of other species, nor is it assimilated by other A particles.

(2) N_A *particles in* V. In a system of N_A particles of A, the expression (4.17) reduces to

$$\mu_A = W(A \mid A) + kT \ln \rho_A \Lambda_A^3 = \{\mu_A^p\} \tag{4.20}$$

where we have the coupling work of A with its surrounding A particles, and the liberation Gibbs energy has changed from $kT \ln(1/V)$ in equation (4.19) to $kT \ln(N_A/V)$ in equation (4.20). This is a typical expression for the CP of a pure gas or liquid. We have also introduced on the rhs of equation (4.20) the conventional notation in thermodynamics, μ_A^p, i.e., the CP of a pure A (sometimes also denoted μ_A°).

As in the previous example, the term $kT \ln \rho_A \Lambda_A^3$ may not be referred to as a mixing term. Clearly, since only A molecules are present there is no "mixing" (in the conventional sense). However, in contrast to the previous example we do have the assimilation term $kT \ln N_A$.

(3) *Single A particle in* N_B *B particles*. Finally, for a system of N_B B molecules and one A molecule, the CP of A in equation (4.17) reduces to

$$\mu_A = W(A \mid B) + kT \ln \rho_A \Lambda_A^3 = \{\mu_A^{\circ\rho} + kT \ln \rho_A\} \tag{4.21}$$

It is noteworthy that the only change that affects the CP in equation (4.19) by the addition of B is the appearance of the coupling work (of A with the surrounding B molecules). The liberation Gibbs energy is *exactly* the same as in equation (4.19) (we note that $\rho_A = 1/V$). Obviously, the mere fact that now A *does mix* with the B molecules does not have any effect on the magnitude or the interpretation of the term $kT \ln \rho_A \Lambda_A^3$ (with $\rho_A = 1/V$). As in example (1), it consists of two contributions — the momentum and volume terms only.

At this point we observe that the mixing process that *did* occur when adding the N_B molecules in this particular example affects only the coupling work. When we treat ideal gases the latter is zero, and the CPs of A in equations (4.19) and (4.21) are identical.

In thermodynamic language the last statement might sound ironic. The change from equation (4.19) to (4.21) due to the *mixing* is realized *only* in the standard CP (μ_A°) and not in what is conventionally referred to as the mixing Gibbs energy (namely $kT \ln \rho_A$)!

A more common misconception in referring to the term $kT \ln \rho_A$ as a mixing term occurs when one transforms from number density to mole fraction. In our example the transformation (assuming $\rho_A \ll \rho_B$) is

$$x_A = \rho_A/\rho_B \tag{4.22}$$

where $\rho_B = N_B/V$. In terms of x_A the CP transforms to

$$\mu_A = W(A \mid B) + kT \ln \rho_B \Lambda_A^3 + kT \ln x_A = \{\mu_A^{ox} + kT \ln x_A\} \quad (4.23)$$

Again, we have written both the statistical mechanical and conventional thermodynamic expressions for the CP.

The reference to $kT \ln x_A$ as the contribution of A to the mixing Gibbs energy (and similarly to $-k \ln x_A$ as the corresponding mixing entropy) is so ubiquitous that it is really difficult to find exceptions. However, the mere fact that we have transformed variables from ρ_A into x_A does not confer any new meaning on $kT \ln x_A$. This term still contains part of the liberation Gibbs energy of A, but in addition it also contains $-kT \ln \rho_B$, which is part of the liberation Gibbs energy of B. Therefore, the term $kT \ln x_A$ cannot be interpreted as the Gibbs energy of mixing.

At this point it should be clear that in the general expression for the CP (4.17), the term $kT \ln \rho_A \Lambda_A^3$, or any transformation of such, has nothing to do with the process of mixing. It has a clearcut meaning as liberation Gibbs energy, independent of whether or not other species are present in the system.

4.4. TWO EXAMPLES OF THERMODYNAMIC NONPROCESSES

In this section we present two process in which an ideal gas expands spontaneously from V to $2V$ in an adiabatic system.

Consider processes III and IV depicted in Figure 4.3. In both, N molecules of ideal gas are initially confined to a volume V. We remove a partition and let the gas expand into a new volume $2V$. In both cases the entropy change is

$$\Delta S^{III} = \Delta S^{IV} = kN \ln 2 > 0 \quad (4.24)$$

From the thermodynamic point of view, these two processes are identical. In both, the driving force for the spontaneous change is the expansion from V to $2V$. This is the *cause* of the positive increase in the entropy and also the source of the irreversibility of the processes.

However, in process III we do observe that some of the gas molecules are elevated, and in process IV some of the gas molecules are transferred to an elongated compartment. These are *observable* phenomena. If we like, we may label the various processes by adding a proper descriptive part. Thus, III may be referred to as the "process of expansion in which we also observe elevation," while IV may be referred to as the "process of expansion in which we also observe elongation." In each of these

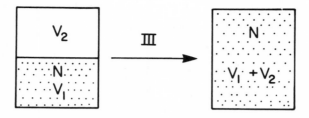

"ELEVATION PROCESS"

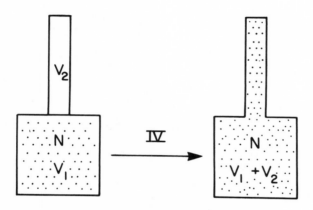

"ELONGATION PROCESS"

Figure 4.3. Two expansion processes. In III, N molecules in volume V_1 are expanded into volume $V_1 + V_2$. The same process takes place in IV, but with slightly modified conditions. The first may be referred to as an "elevation process" and the second as an "elongation process." We assume for simplicity that $V_1 = V_2 = V$ and the process is carried out adiabatically.

sentences the descriptive part ("elevation" or "elongation") might be useful to distinguish between various processes of expansion. However, it would sound ridiculous to refer to ΔS^{III} as the "entropy of elevation" and to ΔS^{IV} as the "entropy of elongation" or to conclude that "the elevation process (or the elongation process) is essentially irreversible."

Why is it so easy for us to reject the terms "entropy of elevation" and "entropy of elongation," yet at the same time almost impossible to repudiate the existence of the "entropy of mixing?"

The answer is quite simple. In the two examples presented in Figure 4.3, we can easily distinguish between the *causative* (expansion) and the *descriptive* (elevation, elongation) parts of the processes. A somewhat

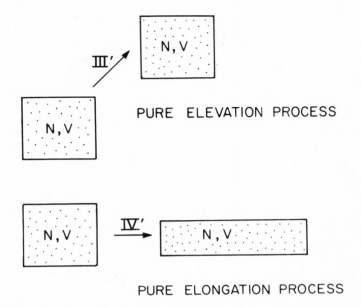

PURE ELEVATION PROCESS

PURE ELONGATION PROCESS

Figure 4.4. Pure processes of "elevation" (no external field is presumed) and "elongation" (no surface effects are presumed).

firmer argument would be to *isolate* the process of elevation and elongation and actually demonstrate that the "pure" processes of elevation and elongation to not change any of the terms in equation (4.18); hence, from the thermodynamic point of view, these are nonprocesses.[†] Schematic descriptions of such "pure elevation" and "pure elongation" are shown in Figure 4.4.

One can easily invent many other expansion processes in which we observe "flow to the right," "flow to the left," "flow downward," and so on. In particular, the last one will be the reverse of process III and, clearly, the "elevation process" as well as its reversed process cannot be "essentially irreversible processes."

We are now in the position to clarify in what sense the term "mixing entropy" is inappropriate and in what sense the statement that the "mixing is essentially irreversible" is incorrect. This becomes almost self-evident if we simply refer to process I as "expansion in which we also

---

[†] Assuming no external fields and no surface effects; otherwise, both "elevation" and "elongation" might become *bona fide* thermodynamic processes, while the "mixing" by itself is always a "nonprocess."

observe mixing." Thus the expansion of the As from V into $2V$ and the expansion of the Bs from V into $2V$ is the *cause* of $\Delta S^I > 0$. The mixing phenomenon is only a *descriptive* part, exactly as the elevation and elongation phenomena.

Perhaps to complete the argument we should also show that there exists a "pure mixing process" and that the reverse of the mixing process can also be carried out spontaneously. The first is easily shown, and in fact it is well known that process II may be carried out reversibly with $\Delta S^{II} = 0$ (see Figure 4.5). If we examine carefully the three terms of the CPs of both A and B during process II, we see that neither the volume accessible to each molecule nor the assimilation terms of A and B are changed throughout the process. Therefore, process II *is* the "pure mixing process" in the same sense as the "pure elevation" and "pure elongation" shown in Figure 4.4. It is "pure" since we observe only mixing, but no other parameters that affect the CP of A and B are changed.

Finally, it is also instructive to demonstrate the existence of a spontaneous demixing process. Figure 4.6 describes such a process. We start with a mixture of A and B in a small volume v. We now let this volume be in contact with two large volumes through a partition which is permeable to A only on one side, and to a second volume through a partition permeable to B only on the other side. The two large volumes are empty initially and the entire system is isolated. If we let the system proceed spontaneously, the entropy change from the initial to the final

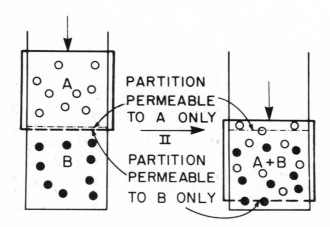

Figure 4.5. A reversible mixing process. This is essentially process II in which we use two partitions permeable to A and B molecules, respectively, to achieve the mixing.

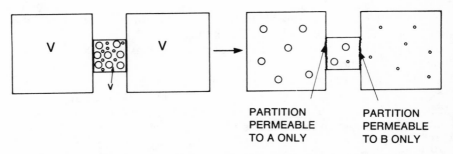

PARTITION
PERMEABLE
TO A ONLY

PARTITION
PERMEABLE
TO B ONLY

Figure 4.6. An irreversible demixing process. A mixture of A and B initially in volume v is brought in contact with two large volumes $V(V \gg v)$ through partitions permeable to A and B, respectively.

state is given by

$$\Delta S = N_A k \ln \left(\frac{V + v}{v}\right) + N_B k \ln \left(\frac{V + v}{v}\right) = 2Nk \ln \left(\frac{V + v}{v}\right) > 0 \quad (4.25)$$

for $N_A = N_B = N$. Clearly, if we choose $V \gg v$, we shall have practically a complete irreversible demixing process. (If we insist on eliminating the remaining small quantity of the mixture in v, we can move the two permeable partitions toward each other quasi-statically. The change in entropy in this step is

$$2Nk \ln \left(\frac{V + \frac{1}{2}v}{V + v}\right) \approx - Nk \frac{v}{V}$$

which can be made negligible for the case $V \gg v$.)

Clearly, the driving force for this irreversible demixing process is exactly the same as the driving force for the irreversible mixing in process I. In both, it is the increase in the accessible volume for each of the molecules.

Had we used the same reasoning as in Section I to conclude that the "mixing process is essentially irreversible," we could equally well conclude that the "demixing process is essentially irreversible." Both of these conclusions are void of thermodynamic content! The "mixing" as well as the "demixing" hold only a descriptive role and, as such, they are, from the thermodynamic point of view, nonprocesses (in the same sense as the "elevation" and "elongation" processes described above).[†]

---

† See the previous footnote.

The conclusion reached above applies strictly to ideal gases. In the case of interacting particles, the interactions produce a new driving force that affects the mixing thermodynamics. This topic will be discussed in Sections 4.9 through 4.11.

At this point, we have completed our proof of the assertions made in Section 4.2. Nevertheless, there still remains the disturbing question of how it happened that concepts such as "mixing entropy" and "mixing Gibbs energy" have gained such a respected (though underserved) place in most textbooks and research articles in thermodynamics. The answer to this queston is not trivial. We shall suggest a possible answer in the next section.

4.5. MACROSCOPIC (MA) AND MICROSCOPIC (MI) VIEWS

In this section we would like to speculate on the origin of the common misconceptions regarding the mixing processes. We believe that these stem from the fact that we "see" different things if we examine the same process by our macroscopic (MA) (or thermodynamic) eyes or by our microscopic (MI) (or molecular) eyes. The following two examples will clarify this point.

Consider first the two processes depicted in Figure 4.7. One is exactly process I of Section 4.2; in the other we have replaced all the B molecules by the A molecules. When we remove the partition from each system, we observed by our MA eyes that in process I' "nothing has

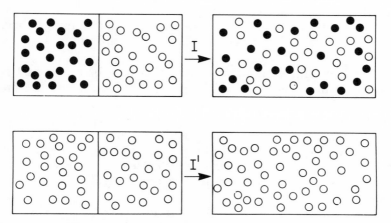

Figure 4.7. Comparison of two "mixing" processes. In I, A and B are distinguishable species; in I', the molecules in the two compartments are identical.

happened"; therefore $\Delta G = \Delta S = 0$, at it should. In process I we *observe* mixing and compute $\Delta G < 0, \Delta S > 0$. This observable difference between process I and process I' led to the conclusion that the *mixing* is *responsible* for the changes in G and S.

The MI view of these processes is quite different. In process I we "see" that the accessible volume of each A and B has increased from V to $2V$. This causes an entropy change of $\Delta S^{I} = 2kN \ln 2$. In process I' we "see" two processes: first, the accessible volume of each molecule changes from V to $2V$, resulting in an entropy change of $2kN \ln 2$; and, in addition, the assimilation terms of each molecule change, resulting in a change of entropy of $-2kN \ln 2$. We see (here in the conventional sense of the word) that process I' looks more complicated to the MI eyes than to the MA eyes. Perhaps it is worthwhile noting that the exact compensation of the expansion and assimilation effects is accidental. Using the statistical mechanical expression for the entropy of ideal gas, we can compute the entropy changes associated with the removal of the partition in processes I and I', namely

$$\Delta S^{I} = 2Nk \ln 2 \tag{4.26}$$

and

$$\Delta S^{I'} = 2Nk \ln 2 - k \ln[(2N)!/(N!)^2] > 0 \tag{4.27}$$

We see that both contain the contribution due to the expansion, $2Nk \ln 2$. The second, however, contains a contribution due to the assimilation process and is always positive, i.e., removing the partition in process I' always causes an increase in the entropy. However, for macroscopic systems, the quantity $\Delta S^{I'}$ is negligibly small and, therefore, practically we say that $\Delta S^{I'} = 0$. The statement, that nothing happens in process I' so that $\Delta S^{I'} = 0$, is clearly an oversimplification of what really happens.

The astonishingly different results obtained for processes I and I' have led to the well-known Gibbs paradox,[69] which results from the fallacy of the assumption that one can pass from distinguishable to indistinguishable molecules continuously. The second consequence of the comparison between processes I and I' is the realization that the two processes are identical except for the *mixing*. Therefore, any difference in the thermodynamic results obtained for these two processes was almost naturally assumed to result from the inherent *irreversibility* of the *mixing process*.

Next, we turn to processes II and II' in Figure 4.8. Looking through our MA eyes, we "see" that process II involves both *mixing* and *compression* (this is where we started in Section 4.2). The MI view is again quite

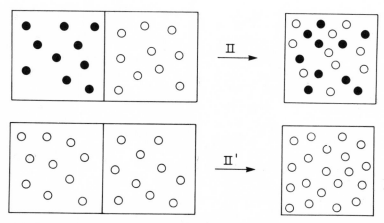

Figure 4.8. Pure mixing II and compression II' processes. In II, A and B are distinguishable species. In II', the molecules are identical.

different. As discussed above in process II, nothing except pure mixing really happens! For process II' we observe a compression process with our MA eyes and compute the entropy change as

$$\Delta S_{\text{Comp}}^{\text{II}'} = 2Nk \ln(V/2V) = 2Nk \ln(1/2) \tag{4.28}$$

In the MI view, however, the volume accessible to each of the molecules did not change. Each A molecule was confined to a volume V before and after the process. However, the assimilation term of each molecule was changed. Each molecule has been assimilated by N molecules before and by $2N$ molecules after the process. The corresponding change in entropy is

$$\Delta S_{\text{Ass}}^{\text{II}'} = k \ln[(N!)^2/(2N)!] \approx 2Nk \ln(1/2) \tag{4.29}$$

which, of course, is *numerically but not conceptually* equal to relation (4.28).

From the above analysis of the difference in the thermodynamics of the two processes I and I', we have concluded that both contain exactly the same contribution from the expansion part (in case I, A and B expand from V into $2V$, and in case I', A from each compartment expands from V into $2V$). Thus the difference in the values of ΔG and ΔS for these two processes does not arise from the conspicuous *mixing* in I but from the inconspicuous *assimilation* in I'.

4.6. IS THERE A "PURE DEASSIMILATION PROCESS?"

The answer to the question posed in this section is not essential to any of the arguments presented in this chapter. We have seen that the assimilation process causes a negative change in entropy [e.g., equation (4.29)] and likewise a positive change in Gibbs energy. This leads us to conclude that the reverse process, which we shall refer to as the *deassimilation* process, should cause an increase in entropy. The natural question that arises is whether there exists a "pure" deassimilation process which occurs spontaneously in an isolated system.

We present here two such processes. The first is essentially a reversal of process II'. Consider $2N$ identical particles confined to a volume V. We remove a partition and let the gas expand from V to $2V$. Next we replace the partition in its original position. The two steps are described schematically in Figure 4.9. To compute the entropy change in the process, we use the ideal-gas expression for the entropy in the initial and final states; the result is:

$$\Delta S = S^f - S^i = 2[kN\ln(V/\Lambda^3) - k\ln N! + \tfrac{3}{2}kN]$$
$$- [2kN\ln(V/\Lambda^3) - k\ln(2N)! + 2\tfrac{3}{2}kN]$$
$$= k\ln[(2N)!/(N!)^2] \approx 2Nk\ln 2 > 0 \qquad (4.30)$$

As discussed in the last part of Section 4.5, in this process the accessible volume of each particle is unchanged in the overall process. The only change that occurs is in the deassimilation of $2N$ particles into two subgroups, each of which contains N particles. In this sense the entire process is a "pure deassimilation" process. However, we have achieved this process by the application of the expansion as the driving force. In the first step the change of entropy due to expansion is $2Nk\ln 2$. In the second step we reduce the volume accessible to each particle: hence a negative contribution of $-2Nk\ln 2$, and at the same time we have a contribution due to the deassimilation in the amount of $2Nk\ln 2$. This is the net amount which is reported in relation (4.30). Thus the entire

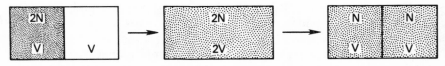

Figure 4.9. A two-step "pure deassimilation" process. First, we let $2N$ particles expand from V to $2V$ by removing a partition, and then the partition is replaced in its original position.

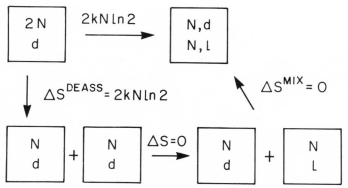

Figure 4.10. A spontaneous process of racemization is performed in three steps. Firs , the system of $2N$ identical particles is deassimilated into two groups, each of which contains N particles. One group of d particles is then transformed into l particles. Finally, the two distinguishable components d and l are mixed. Clearly, the entropy changes only in the deassimilation process in the first step, which is essentially the process described in Figure 4.9.

process is a pure deassimilation process, although it has been coupled with an expansion process.

The second example demonstrates a pure deassimilation process in which the deassimilation is the *only* driving force. Furthermore, it also illustrates how a misconception regarding the "entropy of mixing" has gained legitimacy simply because there was no other way of explaining it.

Consider an ideal gas of $2N$ molecules that contains one chiralic center. Initially, we prepare the system in such a way that all the molecules are in one of the enantiomeric forms, say the d enantiomer. We then introduce a catalyst that induces a racemization process in adiabatic conditions. At equilibrium we obtain N molecules of the d enantiomer and N molecules of the l enantiomer.

The entropy change in this spontaneous process is well known:[70,71]

$$\Delta S = 2kN \ln 2 \tag{4.31}$$

If we analyze carefully the various factors involved in the expression of the CP of the d and l molecules [e.g., equation (4.18)], we find that the momentum partition function as well as any other internal partition function of each molecule, and the volume accessible to each molecule, are unchanged during the entire process. The only change that does take place is in the assimilation terms of d and l; hence the entropy change is due to the deassimilation of $2N$ identical molecules into two subgroups of distinguishable molecules; N of one kind and N of a second kind. This process, along with an alternative route for its realization, is described in Figure 4.10.

This example is also of interest from the following conceptual point of view. The increase of entropy in process I is often ascribed either to a mutual diffusion of one component into the other (induced by the removal of a partition) or to a proper "mixing" (in the sense that two *distinguishable* species are being mixed).

The above example clearly involves *neither* proper mixing *nor* interdiffusion. The spontaneous process that occurs is the transformation of *one* species into *two distinguishable* species, and this process evolves in a homogeneous phase with no apparent mutual diffusion of one species into the other. Therefore, the term "mixing" may not be used here either causatively or even descriptively. (If the process of racemization is referred to as "mixing," should we refer to the reverse of this process as "demixing"?) In spite of this, and curiously enough, the entropy change in this process [e.g., equation (4.31)] is traditionally referred to as the "mixing entropy."[70,71] The only reason that leads to such a misinterpretation is probably the *lack* of recognition of the role of the deassimilation phenomenon. As demonstrated in Figure 4.10, if we let the same process proceed through a different route, the mixing step is exactly the same as in process II and may be carried out reversibly without any change in entropy.

Of course, the deassimilation proccess may also occur in more complex reactions. For example, in a *cis-trans* reaction we have the equilibrium condition

$$N_{cis}/N_{trans} = \exp(-\Delta G^\circ/kT) \tag{4.32}$$

Suppose we start with $2N$ moecules in the *cis* form of dichloroethylene and let the system reach equilibrium, where we have N_{cis} and N_{trans} molecules, and $N_{cis} + N_{trans} = 2N$. In this case the entropy change would have one contribution due to the *reaction* and another contribution due to the *deassimilation* of $2N$ molecules into two subgroups, i.e.,

$$\Delta S = \Delta S_{reaction} + \Delta S_{deassimilation} \tag{4.33}$$

The process of racemization described in Figure 4.10 is a special case of a chemical reaction for which $\Delta G^\circ = 0$; hence $N_d = N_l = N$.

In the examples given in this section we note a kind of hierarchy in the level of the indistinguishability. In the first example the molecules are identical, but they are distinguishable by the mere fact that they occupy different compartments. In the racemization example the molecules have exactly the same internal degrees of freedom, but they differ in their optical activity only. The last example also involves differences in internal degrees of freedom of the molecules. It is of interest to note that in a recent discussion of the Gibbs paradox, Lesk[72] has also suggested viewing ethanol (C_2H_5OH) and dimethyl ether (CH_3OCH_3) as two "isomers," since they share certain highly excited states, namely those in which the molecules are dissociated into the same fragments.

4.7. DELOCALIZATION PROCESS AND THE COMMUNAL ENTROPY

The concept of assimilation may be applied to the so-called delocalization process that has been discussed in connection with the application of lattice theories of fluids.[73]

The process is described schematically in Figure 4.11. The entropy change upon the removal of all the partitions is

$$\Delta S = S^{gas} - S^{loc}$$
$$= kN \ln(V/\Lambda^3) - k \ln N! + \tfrac{3}{2}kN - kN \ln(v/\Lambda^3) - \tfrac{3}{2}kN \quad (4.34)$$

where $v = V/N$ is the volume of each cell. Application of the Stirling approximation yields the well-known result

$$\Delta S = kN \quad (4.35)$$

This is known as the communal entropy. Closer examination reveals that ΔS in equation (4.35) is a net result of two effects: the increase of the accessible volume for each particle from v to V, *and* the assimilation of N

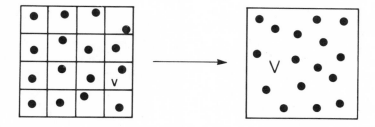

Figure 4.11. Delocalization experiment. N identical particles are initially placed in cells, each of which is of volume v. The partitions between the cells are removed. The entropy change in this process is referred to as the communal entropy.

particles. The two contributions are explicitly

$$\Delta S = kN \ln(V/v) - k \ln N!$$ (4.36)

If we choose $v = V/N$, then the first term is always the dominating one.

Originally, Hirschfelder et al.[73] introduced the concept of communal entropy to explain the entropy of melting of solids. They specifically stated that "this communal sharing of volume gives rise to an entropy of fusion." This idea was later criticized by Rice[74] and by Kirkwood,[75] and now the whole concept of communal entropy in the context of the theory of liquids is considered to be obsolete.[76,79]†

Here we have cited this example to stress the point that the communal entropy in equation (4.35) is not a result of volume change only but a combination of volume change and assimilation.

4.8. CONCLUSION REGARDING IDEAL-GAS MIXTURES

In the previous sections we have examined the suitability of the terms "entropy of mixing" and "Gibbs energy of mixing" as currently applied to the entropy and Gibbs energy of mixing ideal gases. We showed that the term "mixing" may sometimes be used in a descriptive sense but never in a causative sense, i.e., the *mixing* itself is in no way a thermodynamic driving force for the process. It was also shown that the commonly asserted principle, that the mixing process is essentially irreversible, is incorrect. Instead, we have suggested that the process of deassimilation *is* essentially irreversible and *does* provide a new driving force for thermodynamic processes. We now comment on the place of the ideas presented in this chapter in the general context of the second law of thermodynamics.

Consider a general spontaneous process in an isolated system. The second law states that such a spontaneous process involves an increase in entropy. The statistical mechanical interpretation of this law is that any spontaneous process that increases the entropy is equivalent to an increase in the number of accessible states of the system. This is a very general principle, which includes all possible processes of interest in physical chemistry. However, on a somewhat different level we might be interested in classifying and characterizing various sources of entropy change (or increase in number of states), such as expansion, heat transfer,

† It is noteworthy that the concepts of "communal entropy" and "free volume" were discussed in the first edition of *Liquids and Liquid Mixtures* (1959) but omitted from the second (1969) and third (1982) editions.[77]

chemical reaction, and so on. In all of these we remove a constraint in the system that brings about an increase in the number of states. Traditionally, it has been found useful to study and characterize each of these processes separately. Indeed, this is the reason that textbooks on thermodynamics devote special treatments to expansion experiments, chemical reactions, heat transfer, and so on. It is on this level that we also find in most textbooks a section dealing with mixing processes.

The main conclusion of this chapter is that, once we have presented the thermodynamics of *expansion* of ideal gases, there is no need to introduce the mixing process of two *different* gases. The traditional view implies that "mixing" is a new *process* that causes an increase in entropy and therefore should be classified separately from, but on the same level as expansion, chemical reaction, heat transfer, and so on. This is precisely where we believe the concept of "mixing" is misleading. We suggest that the processes of assimilation and deassimilation, though less conspicuous than mixing, should replace the process of mixing in a course on physical chemistry.

4.9. MIXING AND ASSIMILATION IN SYSTEMS OF INTERACTING PARTICLES; ONE-COMPONENT SYSTEMS

In the previous sections of this chapter we have examined the suitability of the terms "entropy of mixing" and "Gibbs energy of mixing" as currently applied to the entropy and Gibbs energy of mixing ideal gases. We showed that the term "mixing" may sometimes be used in a descriptive sense but never in a causative sense, i.e., the *mixing* itself is in no way a thermodynamic driving force for the process.

We now extend our treatment to include systems of interacting particles. It is in these systems that the misconceptions regarding the Gibbs energy of mixing lead not only to incorrect interpretation of experimental results, but actually cause a great deal of waste of "energy." By waste of "energy" we mean the effort expanded in superfluous computations of meaningless quantities. Some examples are cited in Section 4.10.

In the following section we discuss the case of one-component systems of interacting particles. In Section 4.10 we treat the case of two-component systems. These are systems that are of most interest in solution chemistry. In this and the following two sections we shall discuss Gibbs energy changes rather than entropy changes. The reason is that the former is the more commonly discussed quantity in solution thermodynamics. Consider the experiment of removing a partition between two compartments, each of volume V and at the same temperature T, but consisting of different numbers of particles N_1 and N_2 of the same species, say A molecules (Figure 4.12).

The Gibbs energy change for this process (carried out at constant temperature T, total volume $2V$, and total number of particles $N_1 + N_2$) is

$$
\begin{aligned}
\Delta G = G^{\mathrm{f}} - G^{\mathrm{i}} &= N_1 \mu_1^{\mathrm{f}} + N_2 \mu_2^{\mathrm{f}} - N_1 \mu_1^{\mathrm{i}} - N_2 \mu_2^{\mathrm{i}} \\
&= [N_1 W(\mathrm{A} \,|\, \mathrm{f}) - N_1 W(\mathrm{A} \,|\, \mathrm{i}_1) + N_2 W(\mathrm{A} \,|\, \mathrm{f}) - N_2 W(\mathrm{A} \,|\, \mathrm{i}_2)] \\
&\quad + \left[N_1 kT \ln \frac{V}{2V} + N_2 kT \ln \frac{V}{2V} \right] \\
&\quad + \left[N_1 kT \ln \frac{N_1 + N_2}{N_1} + N_2 kT \ln \frac{N_1 + N_2}{N_2} \right]
\end{aligned}
\tag{4.37}
$$

where we have included in brackets three different contributions to the change in the free energy. The first arises from the change in the interaction of the molecules from their initial states (i_1 and i_2) to their common final state (f). The second is the expansion term (for simplicity we choose $V_1 = V_2 = V$). The third term is the contribution due to assimilation. This classification of the various terms is independent of any assumption of ideality of the system. To emphasize the advantage of this point of view, consider the same processes carried out with ideal gases. Here relation (4.37) reduces to

$$
\begin{aligned}
N_1 kT \ln \frac{V}{2V} &+ N_2 kT \ln \frac{V}{2V} + N_1 kT \ln \frac{N_1 + N_2}{N_1} + N_2 kT \ln \frac{N_1 + N_2}{N_2} \\
&= N_1 kT \ln \frac{P_{\mathrm{f}}}{P_{\mathrm{i}_1}} + N_2 kT \ln \frac{P_{\mathrm{f}}}{P_{\mathrm{i}_2}}
\end{aligned}
\tag{4.38}
$$

On the rhs we show how ΔG appears in a typical textbook of thermodynamics, i.e., N_1 molecules are transferred from an initial pressure P_{i_1} into the final pressure P_{f}. Likewise, N_2 molecules are tranferred from P_{i_2} into P_{f}. This way of computing ΔG is valid *only* for ideal gases. Once interactions

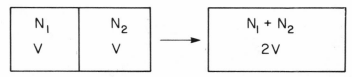

Figure 4.12. Removing a partition separating two compartments of equal volume but different numbers of particles, N_1 and N_2, of the same species.

exist among the particles, the transfer from P_{i_1} to P_f does not tell us anything about the Gibbs energy change. In fact, thermodynamics cannot tell us anything about the significance of the various terms on the lhs of relation (4.38), whether or not the gas is ideal. On the other hand, by introducing the concept of assimilation, we have a clearcut classification of the three terms in relation (4.37), which is valid independent of any assumption of ideality.

Next consider the process depicted in Figure 4.13. Here, again, thermodynamics would have told us that the system is compressed with a Gibbs energy change equal to

$$N_1 kT \ln \left(\frac{P_f}{P_{i_1}}\right) + N_2 kT \ln \left(\frac{P_f}{P_{i_2}}\right) \tag{4.39}$$

This is valid only for ideal gases. Once we have interacting particles, the validity of expression (4.39) as a measure of the Gibbs energy change is lost, and we cannot claim that the compression of the system is the cause of the Gibbs energy change.

Considering the same process from the molecular point of view, we see that each particle is allowed to wander in the *same* volume V, before and after the process. Therefore, here we have only two contributions to ΔG, one due to change in interactions and the second due to assimilation. If the gas is ideal, the only contribution is due to assimilation and amounts to

$$N_1 kT \ln \frac{N_1 + N_2}{N_1} + N_2 kT \ln \frac{N_1 + N_2}{N_2} \tag{4.40}$$

This very term would have been referred to as "compression" in the traditional thermodynamic language.

We now modify this experiment by taking N_A A particles in one box and N_B B particles in the second box. Here again, no volume change is experienced by the particles. Also, no change in the assimilation occurs; hence, if we were treating ideal gases, we would observe no change in

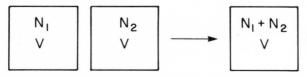

Figure 4.13. A process of transferring two systems of equal volume V, but containing different number of particles, into a single compartment of the same volume V.

Gibbs energy. This case has been discussed in Section 4.2. We stress here that though mixing *is observed*, there is no change in Gibbs energy. The situation is markedly different when we have interactions among the particles. In this case the Gibbs energy change is due only to the change in the interaction terms, namely an A molecule being initially surrounded by As only is surrounded by both As and Bs in the final state. Similarly, the surroundings of B are replaced by a mixture of A and B. It is here that we might legitimately refer to the term "mixing Gibbs energy," since the mixing process has been "noticed" by both A and B through their interaction with their surroundings. We shall elaborate on some specific examples of mixtures of two components of interacting particles in the next section.

4.10. TWO-COMPONENT SYSTEMS OF INTERACTING PARTICLES

Consider a mixture of two components A and B with corresponding number densities ρ_A and ρ_B, respectively. The CPs of the two species are given by

$$\mu_A = W(A \mid A + B) + kT \ln \rho_A \Lambda_A^3 \tag{4.41}$$

and

$$\mu_B = W(B \mid A + B) + kT \ln \rho_B \Lambda_B^3 \tag{4.42}$$

Suppose this mixture was prepared from pure A at initial density ρ_A^p and pure B at initial density ρ_B^p. For simplicity, we assume that the process of mixing was carried out isothermally so that Λ_A^3 and Λ_B^3 do not change upon the process of mixing. The Gibbs energy change for this process, denoted III, is

$$\Delta G^{III} = (N_A \mu_A + N_B \mu_B) - (N_A \mu_A^p + N_B \mu_B^p) \tag{4.43}$$

where the superscript p refers to the "pure" component. Using the more detailed expression for the CPs in equations (4.41) and (4.42), we rewrite equation (4.43) as

$$\Delta G^{III} = \{N_A[W(A \mid A + B) - W(A \mid A)] + N_B[W(B \mid A + B) - W(B \mid B)]\}$$
$$+ [N_A kT \ln(\rho_A/\rho_A^p) + N_B kT \ln(\rho_B/\rho_B^p)] \tag{4.44}$$

This should be compared with relation (4.37). Here, the first term accounts for the changes that take place in the coupling work due to the mixing process. The second term is due to the change in the *volume* accessible to each component upon mixing. Since we started with pure A and B, there is no contribution due to assimilation, a term that we did encounter in relation (4.37).[†] We now turn to some special cases of equation (4.44).

(1) *The symmetrical ideal solutions.* Let A and B be very similar molecules, say isotopes of the same molecule or two molecules having very similar structure and composition. In this case the coupling work of, say, A is determined only by the average number of particles in its surroundings and independent of its composition. This case will be referred to as the symmetrical ideal solution and its experimental manifestation is the well-known Raoult's law.

For simplicity, suppose we start with $N_A = N_B$ and $V_A = V_B = V$. We then let the two components expand into the volume $2V$. This is the same as process I of Section 4.2 but with interacting particles. In this special case the first term on the rhs of equation (4.44) is zero, and the second term reduces to

$$\Delta G^{SI} = N_A kT \ln(1/2) + N_B kT \ln(1/2) \tag{4.45}$$

or more generally to

$$\Delta G^{SI} = N_A kT \ln x_A + N_B kT \ln x_B \tag{4.46}$$

This is very frequently referred to as the Gibbs energy of ideal mixing. However, as we stressed already in Section 4.1, this term arises exactly from the *same* origin as the so-called "Gibbs energy of mixing" of ideal gases, namely from expansion of A and B from V into $2V$, and has nothing to do with the mixing process.

We therefore suggest referral to equations (4.45) and (4.46) simply as the expansion Gibbs energy of A and B, not the mixing free energy. This particular example is important, since it demonstrates that there are real systems of *interacting* particles where the driving force for the process is purely the expansion of each component into a larger volume. A second extreme case is discussed below.

(2) *Mixing involving changes in interactions only.* Let us look at another extreme case (which is the analogue of process II of Section 4.2). Here we start with $N_A = N_B$ and $V_A = V_B$ and also require that the final volume be $V = V_A = V_B$.

[†] We would also encounter assimilation terms had we mixed two mixtures of A and B instead of two *pure* components.

In this case equation (4.44) reduces to

$$\Delta G^{IV} = N_A[W(A \mid A + B) - W(A \mid A)] + N_B[W(B \mid A + B) - W(B \mid B)] \tag{4.47}$$

Here we have no expansion (nor compression) and no assimilation terms. All the Gibbs energy change is due to the changes in the coupling work of A and B. It is perhaps here that the term "mixing Gibbs energy" might be most appropriate to ΔG^{IV}. We note, however, that the conventional "mixing Gibbs energy," a term of the form (4.16), *does not* appear at all. We suggest referring to ΔG^{IV} as the Gibbs energy of *interaction* rather than the *mixing*, to avoid further confusion in this field. We note also that if we "switch off" all the interactions, we recover process II of Section 4.2 for which we have $\Delta G = 0$.

(3) *The dilute-ideal (DI) solutions.* This is the case which has attracted the most attention in solution chemistry. Unfortunately, this is also the case where most of the misconceptions have been involved. Some of these have been discussed elsewhere.[1] Here, we focus only on one aspect of the problem, namely the identification of the so-called mixing terms. It is perhaps here that a cautious examination of the various contributions to the Gibbs energy is important, not only because of its educational aspects, but also because very often excessive and superfluous labor is expanded in computing and interpreting the wrong quantities. As an example, consider the case of a very dilute solution of A in B, such that $N_A \ll N_B$. In this case, when we mix A and B we may assume that

$$W(A \mid A + B) \simeq W(A \mid B) \tag{4.48}$$

$$W(B \mid A + B) \simeq W(B \mid B) \tag{4.49}$$

$$\rho_B = \rho_B^p \tag{4.50}$$

i.e., the addition of a small amount of A to pure B causes a negligible effect on ρ_B^p and on the coupling work of B. In this case the general expression (4.44) reduces to

$$\Delta G^{DI} = N_A[W(A \mid B) - W(A \mid A)] + N_A kT \ln(\rho_A/\rho_A^p) \tag{4.51}$$

Thus a particle A which "sees" only A particles in the pure A is "seeing" now only pure B in its environment. Here, we identify two contributions to the Gibbs energy change: one due to the (extreme) change in the environment of A (from pure A into pure B), and the second to the change of the accessible volume available to A. Neither of these terms is an adequate candidate to be referred to as a mixing term. However, if one insists on using the term "mixing Gibbs energy," it is perhaps more appropriate to refer to the first term on the rhs of equation (4.51), but certainly not to the second term. The traditional view is, quite astonishingly, the opposite. It is very common to refer to the *second* term as the "mixing Gibbs energy." A more conventional form of equation (4.51) is obtained after transforming into mole fractions. In our case the transformation of variables is

$$x_A = \rho_A/(\rho_A + \rho_B) \approx \rho_A/\rho_B^p \qquad (4.52)$$

We now write ΔG^{DI} in the conventional thermodynamic way as

$$\Delta G^{DI} = N_A[\mu_A^{ox} - \mu_B^p] + N_A kT \ln x_A \qquad (4.53)$$

where μ_A^{ox} is the standard CP of A in the mole fraction scale, and μ_A^p is the CP of pure A.

The reference to $N_A kT \ln x_A$ as a "mixing Gibbs energy" term is so ubiquitous that it is really difficult to find exceptions. This erroneous interpretation of the term $kT \ln x_A$ also leads to an erroneous interpretation of the term $\mu_A^{ox} - \mu_A^p$, which is very common in solution chemistry. Beyond the interpretive aspect of this topic, there is also a more important aspect to it, and the following example will demonstrate this point.

When investigating dilute solutions of, say, argon in water or xenon in organic solvents, sometimes the stated purpose of the investigators is to study the Gibbs energy of interaction (or solvation), the entropy of interaction, and so on, of the solute in the solvent. Usually, the primary quantity that is obtained in the experiment is a partition coefficient, i.e., a ratio of the number densities of the solute between two phases. This ratio is directly related to the quantity $W(A \mid A + B)$, which measures the solvation Gibbs energy of the solute A in the solvent.

However, because of the prevalent erroneous interpretation of the term $RT \ln x_A$ as a "mixing term" (sometimes referred to as a "cratic" term[3,78]), one takes the trouble to transform the concentration units from number densities into mole fractions and then compute a so-called "standard Gibbs energy of solution." This is also referred to as a "unitary" term.[3,78] Unfortunatey, the whole procedure involves superfluous work and, in fact, misses the principal aim of the investigation. As we have seen, $kT \ln x_A$ is *not* a mixing Gibbs energy. Furthermore, the quantity left after subtracting this term from ΔG^{DI} does not necessarily have the meaning so often assigned to it.

It is ironic to observe that the primary quantity that one measures experimentally is directly related to the solvation Gibbs energy $W(A \mid B)$.

Instead, a procedure is undertaken to convert the raw data into a less meaningful quantity — an effort that could be saved if the origin of the various contributions to ΔG^{DI} would have been carefully examined. Some specific examples have been critically discussed elsewhere.[18,39]

4.11. TWO BAFFLING EXPERIMENTS INVOLVING A "MIXING GIBBS ENERGY" OF $-2NkT \ln 2$

We present here two "thought experiments" (which in principle could be carried out in practice if the right solutes and solvents, as specified below, could be found) that serve to demonstrate the moral of this chapter: what we observe with our eyes should not, in general, be trusted for reaching thermodynamic conclusions.

We recall processes I and I' of Figure 4.7. In I we remove a partition that separates two distinguishable components of ideal gases. As is well known, the Gibbs energy change is

$$\Delta G^{I} = -2kTN \ln 2 \qquad (4.54)$$

where N is the number of molecules in each compartment. In process I' where we have the same component in the two compartments, the Gibbs energy change is

$$\Delta G^{I'} = 0 \qquad (4.55)$$

We now modify these experiments and show that in an apparently "no-mixing" and "demixing" process we obtain Gibbs energy changes which are *exactly* equal to the quantity (4.54).

To demonstrate these cases, consider again processes I and I' (Figure 4.7). Suppose A and B are colored molecules — say A is blue and B is yellow. Assume also that $N_A = N_B = N$ and the systems are dilute enough so that the gases are ideal. We carry out the two experiments in front of a class of students under a constant temperature. Removing the partitions in processes I and I' leads to conclusions, which we state as follows:

1. In process I we *observe mixing*. The entire system becomes green, and we calculate $\Delta G^{I} = -2NkT \ln 2$.
2. In process I' we *observe no mixing*. The entire system remains blue, and we calculate $\Delta G^{I'} = 0$.

After removing the partitions, we reach an equilibrium state. At equilibrium, we can introduce the partitions back into their original places or remove them with no discernible effect on the system and no change in Gibbs energy.

Next we place the two partitions back in their original places and modify the experimental situation as follows (Figure 4.14).

We fill the two compartments with two different liquids that are transparent to visible light. We also assume that we have chosen the solutes A and B in such a way that their colors are not affected by the presence of the solvents. We require also that the two liquids be completely immiscible with each other. (We may also use miscible solvents, but then we should require that the partitions be permeable to A and B only and not to the solvents.)

We now remove the partitions, the audience being unaware of the presence of the solvents in the two compartments. For them, the system looks exactly the same as in the original experiment. We claim that now the removal of the partition has caused the following results [compare with (1) and (2) above]:

1. In process I* we *observe demixing*. The entire green system separates into yellow and blue. We claim that

$$\Delta G^{I*} = - 2NkT \ln 2$$

2. In process I*' we *observe no mixing*. The entire system remains blue, and we claim that $\Delta G^{I*'} = - 2NkT \ln 2$

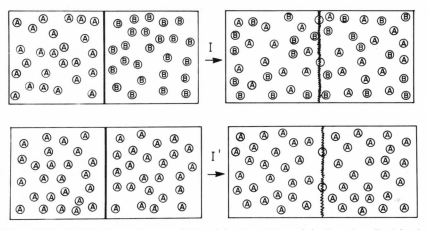

Figure 4.14. Modified experiments of "demixing" and "no-mixing" as described in the text.

The results given above are correct. It is also true that removing the partitions causes only the flow of A or A and B between the two compartments, but not of the solvents (which are presumed to be immiscible).

Clearly, anyone who observes these experiments must be puzzled if he is unaware of the presence of the two liquids. For him the compartments still contain ideal gases, and therefore processes I* and I*′ are expected to yield zero Gibbs energy changes.

Evidently, the trick was made possible by introducing a new driving force into the system — the difference in the solvation Gibbs energies of A and B in the two liquids. The existence of these particular results for ΔG^{I*} and $\Delta G^{I*'}$ may be proved. A brief outline of the proof and choice of parameters is outlined in the footnote.[†] We believe that following through this exercise will be rewarding, since it demonstrates that driving forces may not be visible, and a visible phenomenon may not be the driving force. In the words of Isaiah[79]: "And he shall not judge after the sight of his eyes, neither reprove after the hearing of his ears."

[†] In experiment I*′, let $N_A^\alpha = N_A^\beta$ be the initial concentrations of A in the two phases α and β. We can choose the solvation Gibbs energies of A in these two phases in such a way that the partition coefficient K_A, defined by

$$K_A = \left(\frac{\bar{N}_A^\beta}{\bar{N}_A^\alpha}\right)_{eq} = \exp\left[\frac{W(A\,|\,\alpha) - W(A\,|\,\beta)}{kT}\right]$$

fulfills the condition

$$\frac{\Delta G^{I*'}}{kT(N_A^\alpha + N_A^\beta)} = \ln 2 + \ln\left(\frac{\sqrt{K_A}}{1 + K_A}\right) = -\ln 2$$

A choice of $K_A = 12.276$ or $K_A = (12.276)^{-1}$ is a solution of this equation.

For the "demixing" experiment I*, we start with a homogeneous distribution of A and B in the system. We define K_A and K_B as above and require that the Gibbs energy change be

$$\Delta G^{I*} = -kT(N_A + N_B)\ln 2$$

where N_A and N_B are the total number of A and B molecules in the system. A choice of $K_A = 12.276$ or $K_B = (12.276)^{-1}$ will lead to an almost complete demixing of A and B (about 93% of As in one compartment and 93% of Bs in the other).

Tables of Some Physical Constants and Conversions

Table I
The System of the SI Base Units

Concept	Symbol	Name of unit	Abbreviation
Length	l	meter	m
Mass	m	kilogram	kg
Time	t	second	s
Electric current	I	ampere	A
Thermodynamic temperature	T	kelvin	K
Amount of substance	n	mole	mol

Table II
Derived SI Units

Concept	Name of unit	Abbreviation of unit name	Unit in other SI units	Units in SI base units
Frequency	Hertz	Hz		s^{-1}
Force	Newton	N		$kg\ m\ s^{-2}$
Pressure	Pascal	Pa	N/m^2	$kg\ m^{-1}\ s^{-2}$
Energy	Joule	J	Nm	$kg\ m^2\ s^{-2}$
Power	Watt	W	J/s	$kg\ m^2\ s^{-3}$
Electric charge	Coulomb	C		$A\ s$
Electric potential	Volt	V	W/A	$kg\ m^2\ s^{-3}\ A^{-1}$
Capacitance	Farad	F	C/V	$m^{-2}\ kg^{-1}\ s^4\ A^2$
Electric resistance	Ohm			V/A	$m^2\ kg\ s^{-3}\ A^{-2}$
Conductance	Siemens	S	A/V	$m^{-2}\ kg^{-1}\ s^3\ A^2$

Table III
Prefixes of SI Units

Fraction	Prefix	Symbol	Multiple	Prefix	Symbol
10^{-1}	deci	d	10	deca	da
10^{-2}	centi	c	10^2	hecto	h
10^{-3}	milli	m	10^3	kilo	k
10^{-6}	micro	μ	10^6	mega	M
10^{-9}	nano	n	10^9	giga	G
10^{-12}	pico	p	10^{12}	tera	T
10^{-15}	femto	f	10^{15}	peta	P
10^{-18}	atto	a	10^{18}	exa	E

Table IV
Physical Constants

Quantity	Symbol	Value (SI)	SI unit	Other units
Atomic mass unit	u	1.66056×10^{-27}	kg	
Avogadro constant	N_A	6.0220×10^{23}	mol^{-1}	
Bohr magneton	μ_B	9.2740×10^{-24}	$J\,T^{-1}$	
Bohr radius	a_0	5.29177×10^{-11}	m	
Boltzmann constant	k	1.3807×10^{-23}	$J\,K^{-1}$	
Permittivity of vacuum	ε_0	8.854188×10^{-12}	$F\,m^{-1}$	
Electron charge	e	1.60219×10^{-19}	C	4.80286×10^{-10} esu
Faraday constant	F	9.64846×10^4	$C\,mol^{-1}$	
Nuclear magneton	μ_N	5.0508×10^{-27}	$J\,T^{-1}$	
Speed of light in vacuum	c	2.99792×10^8	$m\,s^{-1}$	
Permeability of vacuum	μ_0	1.2566370×10^{-6}	$H\,m^{-1}$	
Molar gas constant	R	8.3144	$J\,mol^{-1}\,K^{-1}$	$1.987\ cal\ K^{-1}mol^{-1}$
Molar volume, ideal gas ($T = 273.15$ K, $p = 1$ atm)	V_0	22.4138×10^{-3}	$m^3\,mol^{-1}$	
Planck constant	h	6.6261×10^{-34}	J s	
Electron rest mass	m_e	9.1095×10^{-31}	kg	
Proton rest mass	m_p	1.67264×10^{-27}	kg	
Rydberg constant	R_∞	1.097373×10^7	m^{-1}	

Table V
Some Conversion Factors

Concept	Name of unit	SI
Length	Angstrom	10^{-10} m
Volume	Liter	10^{-3} m^3
Force	Dyn	10^{-5} N
	Kilogram force	9.806 N
Pressure	Atmosphere	101325 Pa
	Torr = mm Hg	133.32 Pa
	Bar = 0.9869 atm	10^5 Pa
Viscosity	Poise	0.1 Pas
Energy	Calorie	4.184 J
	Electron volt	1.62022×10^{-19} J
	Erg	10^{-7} J
Temperature	0 °C	273.15 K

List of Abbreviations

CN	coordination number
CP	chemical potential
DI	dilute ideal
HB	hydrogen bond
HI	hydrophobic interaction
HS	hard sphere
ig	ideal gas
lhs	left-hand side
MA	macroscopic
MI	microscopic
PCP	pseudochemical potential
rhs	right-hand side
SI	symmetrical ideal
SOW	structure of water
SPT	scaled-particle theory

References

1. A. Ben-Naim, *J. Phys. Chem.* **82**, 792 (1978)
2. *Encylopaedia Britannica* (15th edn.), Vol. 16, p. 1055 (1974).
3. R. W. Gurney, *Ionic Processes in Solution*, Dover Publishing, New York (1953).
4. A. Ben-Naim, *Water and Aqueous Solutions*, Plenum Press, New York (1974).
5. J. S. Rowlinson and F. L. Swinton, *Liquids and Liquid Mixtures* (3rd edn.), Butterworths Scientific, London (1982).
6. J. G. Kirkwood and F. P. Buff, *J. Chem. Phys.* **19**, 774 (1951).
7. G. A. Cook (ed.), *Argon, Helium and the Rare Gases*, Vol. I, Interscience Publishers, New York (1961).
8. V. G. Fastovskii, A. E. Rovinskii, and Yu. V. Petrovskii, *Inert Gases* (translated from Russian by J. Schmorak), Cat. No. 1876, IPST, Jerusalem (1967).
9. F. Theeuwes and R. J. Bearman, *J. Chem. Thermodyn.* **2**, 171, 179, 501, 507, 513 (1970).
10. J. O. Hirschfelder, C. F. Curtiss, and R. B. Bird, *Molecular Theory of Gases and Liquids* (2nd edn.), John Wiley, New York (1963).
11. A. Ben-Naim, *J. Phys. Chem.* **89**, 5738 (1985).
12. J. H. Hildebrand, *J. Am. Chem. Soc.* **37**, 970 (1915).
13. R. H. Davies, A. G. Duncan, G. Saville, and L. A. K. Staveley, *Trans. Faraday Soc.* **63**, 855 (1967).
14. C. H. Chui and F. B. Canfield, *Trans. Faraday Soc.* **67**, 2933 (1971).
15. J. C. G. Calado and L. A. K. Staveley, *Trans. Faraday Soc.* **67**, 289 (1971).
16. G. L. Pollack, *J. Chem. Phys.* **75**, 5875 (1981).
17. G. L. Pollack and J. F. Himm, *J. Chem. Phys.* **77**, 3221 (1982).
18. A. Ben-Naim and Y. Marcus, *J. Chem. Phys.* **80**, 4438 (1984).
19. G. L. Pollack, J. F. Himm, and J. J. Enyeart, *J. Chem. Phys.* **81**, 3239 (1984).
20. R. Battino and H. L. Clever, *Chem. Rev.* **66**, 395 (1966).
21. E. Wilhelm, R. Battino, and R. J. Wilcock, *Chem. Rev.* **77**, 219 (1977).
22. E. Wilhelm and R. Battino, *Chem. Rev.* **73**, 1 (1973).
23. R. Battino, *Fluid Phase Equilibria* **15**, 231 (1984).
24. F. Franks, in: *Water: A Comprehensive Treatise* (F. Franks, ed.), Vol. 2, Plenum Press, New York (1973).
25. D. D. Eley, *Trans. Faraday Soc.* **35**, 1281 (1939).
26. D. D. Eley, *Trans. Faraday Soc.* **40**, 184 (1944).
27. R. Crovetto, R. Fernandez-Prini, and M. L. Japas, *J. Chem. Phys.* **76**, 1077 (1982).
28. H. S. Frank and M. W. Evans, *J. Chem. Phys.* **13**, 507 (1945).

29. A. Ben-Naim, *J. Chem. Phys.* **42**, 1512 (1965).
30. A. Ben-Naim, *Hydrophobic Interactions*, Plenum Press, New York (1980).
31. A. Ben-Naim, *J. Phys. Chem.* **79**, 1268 (1975).
32. Y. Marcus and A. Ben-Naim, *J. Chem. Phys.* **83**, 4744 (1985).
33. A. Ben-Naim, *J. Chem. Phys.* **82**, 4662, 4670 (1985).
34. R. A. Robinson and R. H. Stokes, *Electrolyte Solutions*, Butterworths Scientific, London (1959).
35. K. N. Marsh and A. E. Richards, *Aust. J. Chem.* **33**, 2121 (1980).
36. J. A. Larkin, *J. Chem. Thermodyn.* **7**, 137 (1975).
37. R. C. Pemberton and C. J. Mash, *J. Chem. Thermodyn.* **10**, 867 (1978).
38. A. Ben-Naim, *J. Chem. Phys.* **82**, 4668 (1985).
39. A. Ben-Naim and Y. Marcus, *J. Chem. Phys.* **81**, 2016 (1984).
40. C. McAuliffe, *J. Phys. Chem.* **70**, 1267 (1966).
41. T. L. Hill, *Introduction to Statistical Thermodynamics*, Addison Wesley, Reading, Massachusetts (1960).
42. K. P. Huber and G. Herzberg, *Molecular Spectra and Molecular Structure*, *Vol. IV*, *Constants of Diatomic Molecules*, Van Nostrand Reinhold Company, New York (1978).
43. S. J. Bates and H. D. Kirschman, *J. Am. Chem. Soc.* **41**, 1991 (1919).
44. R. Haase, H. Naas, and H. Thumm, *Z. Phys. Chem.* **37**, 210 (1963).
45. *The NBS Tables of Chemical Thermodynamical Properties*, Journal of Physical and Chemical Reference Data **11**, 1892 supplement No. 2, NBS, Washington (1982).
46. A. Ben-Naim, *J. Phys. Chem.* **89**, 3791 (1985).
47. V. P. Shanbhag and C. G. Axelsson, *Eur. J. Biochem.* **60**, 17 (1975).
48. A. Ben-Naim, Hydrophobic interactions in biological systems, Chapter 1, in: *Topics in Molecular Pharmacology* (G. C. K. Roberts and A. Burgen, eds.), Elsevier North-Holland Biochemical Press, New York (1982).
49. T. L. Hill, *Statistical Mechanics*, McGraw-Hill, New York (1956).
50. H. Reiss, *Adv. Chem. Phys.* **9**, 1 (1966).
51. T. L. Hill, *J. Chem. Phys.* **28**, 1179 (1958).
52. H. Reiss, H. L. Frisch, and J. L. Lebowitz, *J. Chem. Phys.* **31**, 369 (1959).
53. H. Reiss, H. L. Frisch, E. Helfand, and J. L. Lebowitz, *J. Chem. Phys.* **32**, 119 (1960).
54. A. Ben-Naim, *J. Phys. Chem.* **82**, 874 (1978).
55. A. Ben-Naim, *Faraday Symposium No. 17*, Reading (1982).
56. A. Ben-Naim, *J. Chem. Phys.* **63**, 2064 (1975).
57. A. Ben-Naim, *J. Chem . Phys.* **67**, 4884 (1977).
58. E. Matteoli and L. Lepori, *J. Chem. Phys.* **80**, 2856 (1984).
59. Y. Marcus, *Ion Solvation*, Wiley–Interscience, New York (1985).
60. E. Grunwald, *J. Am. Chem. Soc.* **82**, 5801 (1960).
61. E. Grunwald, in: *Electrolytes*, Pergamon Press, New York (1962).
62. M. Born, *Z. Phys.* **1**, 45 (1920).
63. M. Yaacobi and A. Ben-Naim, *J. Phys. Chem.* **78**, 175 (1974).
64. A. Ben-Naim, *J. Chem. Phys.* **54**, 1387 (1971).
65. A. Ben-Naim and J. Wilf, *J. Chem. Phys.* **70**, 771 (1979).
66. K. Denbigh, *The Principles of Chemical Equilibrium*, Cambridge University Press, London (1966),
67. M. Planck, *Treatise on Thermodynamics* (translated by A. Ogg), Dover Publications, New York (1926).
68. F. H. MacDougall, *Thermodynamics and Chemistry*, John Wiley, New York (1948).
69. J. W. Gibbs, *Scientific Papers*, *Vol. I*, *Thermodynamics*, p. 166, Longmans Green and Co., New York (1906).

70. E. L. Eliel, *Stereochemistry of Carbon Compounds*, McGraw-Hill, New York (1962).
71. J. Jacques, A. Collet, and S. H. Wilen, *Enantiomers, Racemates and Resolutions*, John Wiley, New York (1981).
72. A. M. Lesk, *J. Phys. A* **13**, L 11 (1980).
73. J. O. Hirschfelder, D. Stevenson, and H. Eyring, *J. Chem. Phys.* **5**, 896 (1937).
74. O. K. Rice, *J. Chem. Phys.* **6**, 476 (1938).
75. J. G. Kirkwood, *J. Chem. Phys.* **8**, 380 (1950).
76. A. Münster, *Statistical Thermodynamics*, Vol. 2, Springer-Verlag, Berlin (1974).
77. J. S. Rowlinson, *Liquids and Liquid Mixtures*, Butterworths Scientific, London (1959).
78. C. Tanford, *The Hydrophobic Effect*, Wiley–Interscience, New York (1973).
79. Isaiah 11:3.

Index

RANDALL LIBRARY-UNCW

3 0490 0088782 6